T0181335

Communications in Computer and Information Science 1326

More information about this series at http://www.springer.com/series/7899

Tim Verbelen · Pablo Lanillos ·
Christopher L. Buckley · Cedric De Boom (Eds.)

Active Inference

First International Workshop, IWAI 2020
Co-located with ECML/PKDD 2020
Ghent, Belgium, September 14, 2020
Proceedings

 Springer

Editors
Tim Verbelen (iD)
Ghent University
Ghent, Belgium

Christopher L. Buckley (iD)
University of Sussex
Brighton, UK

Pablo Lanillos (iD)
Donders Institute for Brain, Cognition
and Behaviour
Nijmegen, The Netherlands

Cedric De Boom (iD)
Ghent University
Ghent, Belgium

ISSN 1865-0929   ISSN 1865-0937   (electronic)
Communications in Computer and Information Science
ISBN 978-3-030-64918-0   ISBN 978-3-030-64919-7   (eBook)
https://doi.org/10.1007/978-3-030-64919-7

This Springer imprint is published by the registered company Springer Nature Switzerland AG
The registered company address is: Gewerbestrasse 11, 6330 Cham, Switzerland

Preface

Active inference is a theory of behavior and learning that originated in neuroscience. The basic assumption is that intelligent agents entertain a generative model of their environment, and their main goal is to minimize surprise or, more formally, their free energy. Active inference not only offers an interesting framework for understanding behavior and the brain, but also to develop artificial intelligent agents and to investigate novel machine learning algorithms.

This volume presents some recent developments in active inference and its applications. These papers were presented and discussed at the First International Workshop on Active Inference (IWAI 2020), which was held in conjunction with the European Conference on Machine Learning and Principles and Practice of Knowledge Discovery in Databases (ECML-PKDD) in Ghent, Belgium. The workshop took place fully online on September 14, 2020, due to the COVID-19 pandemic. Out of 25 submissions, 13 full papers and 6 poster papers were selected through a double-blind review process.

The papers are clustered in three sections. The first section is on the application of active inference for continuous system control. The second section bundles papers on the cross-section between active inference and machine learning. Finally, the third section presents some work on active inference in biology and more theoretical aspects.

The IWAI 2020 organizers would like to thank the Program Committee for their valuable review work, all authors for their contributions, all attendees for the fruitful discussions, and Philipp Schwartenbeck, Sindy Löwe, and Rosalyn Moran for their outstanding invited talks.

September 2020

Tim Verbelen
Pablo Lanillos
Christopher L. Buckley
Cedric De Boom

Organization

Organizing Committee

Christopher L. Buckley	University of Sussex, UK
Cedric De Boom	Ghent University - imec, Belgium
Pablo Lanillos	Donders Institute, The Netherlands
Tim Verbelen	Ghent University - imec, Belgium

Program Committee

Glen Berseth	University of California, Berkeley, USA
Christopher L. Buckley	University of Sussex, UK
Cedric De Boom	Ghent University - imec, Belgium
Bart Dhoedt	Ghent University - imec, Belgium
Karl Friston	University College London, UK
Casper Hesp	University of Amsterdam, The Netherlands
Pablo Lanillos	Donders Institute, The Netherlands
Christoph Mathys	Aarhus University, Denmark
Rosalyn Moran	King's College London, UK
Ayca Ozcelikkale	Uppsala University, Sweden
Maxwell Ramstead	McGill University, Canada
Noor Sajid	University College London, UK
Philipp Schwartenbeck	University College London, UK
Alexander Tschantz	University of Sussex, UK
Kai Ueltzhöffer	Heidelberg University, Germany
Tim Verbelen	Ghent University - imec, Belgium
Martijn Wisse	Delft University of Technology, The Netherlands

Contents

Active Inference: Theory and Biology

Active Inference and Continuous Control

On the Relationship Between Active Inference and Control as Inference

Beren Millidge[1(✉)], Alexander Tschantz[2,3], Anil K. Seth[2,3,4], and Christopher L. Buckley[3]

[1] School of Informatics, University of Edinburgh, Edinburgh, UK
beren@millidge.name
[2] Sackler Center for Consciousness Science, Brighton, UK
[3] Evolutionary and Adaptive Systems Research Group, University of Sussex, Brighton, UK
[4] CIFAR Program on Brain, Mind, and Consciousness, Toronto, Canada

Abstract. Active Inference (AIF) is an emerging framework in the brain sciences which suggests that biological agents act to minimise a variational bound on model evidence. Control-as-Inference (CAI) is a framework within reinforcement learning which casts decision making as a variational inference problem. While these frameworks both consider action selection through the lens of variational inference, their relationship remains unclear. Here, we provide a formal comparison between them and demonstrate that the primary difference arises from how the notion of rewards, goals, or desires is incorporated into their generative models. We highlight how the encoding of value leads to subtle differences in the respective objective functionals and discuss how these distinctions lead to different exploratory behaviours.

1 Introduction

Active Inference (AIF) is an emerging framework from theoretical neuroscience which proposes a unified account of perception, learning, and action [11,13–15]. This framework posits that agents embody a generative model of their environment and that perception and learning take place through a process of variational inference on this generative model, achieved by minimizing an information-theoretic quantity – the variational free energy [5,11,16,36]. Moreover, AIF argues that action selection can also be cast as a process of inference, underwritten by the same mechanisms which perform perceptual inference and learning. Implementations of this framework have a degree of biological plausibility [37] and are supported by considerable empirical evidence [12,34]. Recent work has demonstrated that active inference can be applied to high-dimensional tasks and environments [10,21,22,25,29,32,33,35].

The field of reinforcement learning (RL) is also concerned with understanding adaptive action selection. In RL, agents look to maximise the expected sum of rewards. In recent years, the framework of control as inference (CAI) [1,2,7, 20,26,28] has recast the problem of RL in the language of variational inference, generalising and contextualising earlier work in stochastic optimal control on the general duality between control and inference under certain conditions [6, 30, 31]. Instead of maximizing rewards, agents must infer actions that lead to the optimal trajectory. This reformulation enables the use of powerful inference algorithms in RL, as well as providing a natural method for promoting exploratory behaviour [1, 17, 18].

Both AIF and CAI view adaptive action selection as a problem of inference. However, despite these similarities, the formal relationship between the two frameworks remains unclear. In this work, we attempt to shed light on this relationship. We present both AIF and CAI in a common language, highlighting connections between them which may have otherwise been overlooked. We then move on to consider the key distinction between the frameworks, namely, how 'value', 'goals' or 'desires' are encoded into the generative model. We discuss how this distinction leads to subtle differences in the objectives that both schemes optimize, and suggest how these differences may impact behaviour.

2   Formalism

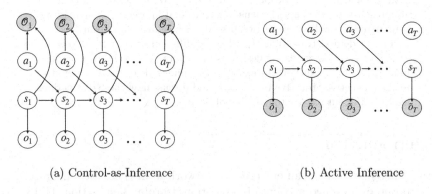

(a) Control-as-Inference   (b) Active Inference

Fig. 1. Graphical models for CAI and AIF. CAI augments the standard POMDP structure with biased (grey-shaded) optimality variables $\mathcal{O}_{t:T}$. AIF simply biases the observation nodes of the POMDP directly.

Both AIF and CAI can be formalised in the context of a partially observed Markov Decision Process (POMDP). Let $\mathbf{a}$ denote actions, $\mathbf{s}$ denote states and $\mathbf{o}$ denote observations. In a POMDP setting, state transitions are governed by $\mathbf{s}_{t+1} \sim p_{\text{env}}(\mathbf{s}_{t+1}|\mathbf{s}_t, \mathbf{a}_t)$ whereas observations are governed by $\mathbf{o}_t \sim p_{\text{env}}(\mathbf{o}_t|\mathbf{s}_t)$. We also assume that the environment possess a reward function $r : \mathcal{S} \times \mathcal{A} \to \mathbb{R}^1$ which maps from state-action pairs to a scalar reward. Agents encode (and potentially learn) a generative model $p(\mathbf{s}_{t:T}, \mathbf{a}_{t:T}, \mathbf{o}_{t:T})$ that describes the relationship between states, actions, observations. AIF and CAI are both concerned with inferring the posterior distribution over latent variables $p(\mathbf{a}_{t:T}, \mathbf{s}_{t:T}|\mathbf{o}_{t:T})$.

However, solving this 'value-free' inference problem will not lead to adaptive behaviour. Instead, some additional assumptions are required to bias inference towards inferring actions that lead to 'valuable' states (Fig. 1).

3   Control as Inference

CAI incorporates the notion of value by introducing an additional 'optimality' variable $\mathcal{O}_t$, where $\mathcal{O}_t = 1$ implies that time step t was optimal. In what follows, we simplify notation by assuming $p(\mathcal{O}_t) := p(\mathcal{O}_t = 1)$. The goal of CAI is then to recover the posterior over states and actions, given the belief that the agent will observe itself being optimal, i.e. $p(\mathbf{s}_t, \mathbf{a}_t | \mathbf{o}_t, \mathcal{O}_t)$. By including the optimality variable we can write the agent's generative model as $p(\mathbf{s}_t, \mathbf{a}_t, \mathbf{o}_t, \mathcal{O}_t) = p(\mathcal{O}_t | \mathbf{s}_t, \mathbf{a}_t) p(\mathbf{o}_t | \mathbf{s}_t) p(\mathbf{a}_t | \mathbf{s}_t) p(\mathbf{s}_t | \mathbf{s}_{t-1}, \mathbf{a}_{t-1})$[1]. Inferring the posterior $p(\mathbf{s}_t, \mathbf{a}_t | \mathbf{o}_t, \mathcal{O}_t)$ is generally intractable, but it can approximated by introducing an auxillary distribution $q_\phi(\mathbf{s}_t, \mathbf{a}_t) = q_\phi(\mathbf{a}_t | \mathbf{s}_t) q(\mathbf{s}_t)$, where ϕ are the parameters of the variational policy distribution $q_\phi(\mathbf{a}_t | \mathbf{s}_t)$, and then optimising the variational bound $\mathcal{L}(\phi)$:

$$\mathcal{L}(\phi) = D_{\mathrm{KL}}\Big(q_\phi(\mathbf{s}_t, \mathbf{a}_t) \| p(\mathbf{s}_t, \mathbf{a}_t, \mathbf{o}_t, \mathcal{O}_t)\Big)$$

$$= \underbrace{-\mathbb{E}_{q_\phi(\mathbf{s}_t, \mathbf{a}_t)}\big[\ln p(\mathcal{O}_t | \mathbf{s}_t, \mathbf{a}_t)\big]}_{\text{Extrinsic Value}} + \underbrace{D_{\mathrm{KL}}\Big(q(\mathbf{s}_t) \| p(\mathbf{s}_t | \mathbf{s}_{t-1}, \mathbf{a}_{t-1})\Big)}_{\text{State divergence}} \tag{1}$$

$$+ \underbrace{\mathbb{E}_{q(\mathbf{s}_t)}\big[D_{\mathrm{KL}}\big(q_\phi(\mathbf{a}_t | \mathbf{s}_t) \| p(\mathbf{a}_t | \mathbf{s}_t)\big)\big]}_{\text{Action Divergence}} - \underbrace{\mathbb{E}_{q_\phi(\mathbf{s}_t, \mathbf{a}_t)}\big[\ln p(\mathbf{o}_t | \mathbf{s}_t)\big]}_{\text{Observation Ambiguity}}$$

where D_{KL} is a Kullback-Leibler divergence. Minimising Eq. 1 – a process known as variational inference – will cause the approximate posterior $q_\phi(\mathbf{s}_t, \mathbf{a}_t)$ to tend towards the true posterior $p(\mathbf{s}_t, \mathbf{a}_t | \mathbf{o}_t, \mathcal{O}_t)$, and cause the variational free energy to approach the marginal-likelihood of optimality $p(\mathcal{O}_t)$.

The second equality in Eq. 1 demonstrates that this variational bound can be decomposed into four terms. The first term (extrinsic value) quantifies the likelihood that some state-action pair is optimal. In the CAI literature, the likelihood of optimality is usually defined as $p(\mathcal{O}_t | \mathbf{s}_t, \mathbf{a}_t) := e^{r(\mathbf{s}_t, \mathbf{a}_t)}$, such that $\ln p(\mathcal{O}_t | \mathbf{s}_t, \mathbf{a}_t) = r(\mathbf{s}_t, \mathbf{a}_t)$. Extrinsic value thus quantifies the expected reward of some state-action pair, such that minimising $\mathcal{L}(\phi)$ maximises expected reward. The state divergence and action divergence terms quantify the degree to which beliefs about states and actions diverge from their respective priors. The approximate posterior over states and the agent's model of state dynamics are assumed to be equal $q(\mathbf{s}_t) := p(\mathbf{s}_t | \mathbf{s}_{t-1}, \mathbf{a}_{t-1})$, such that the agent believes it has no control over the dynamics except through action. This assumption eliminates the

---

[1] Note that CAI is usually formulated in the context of an MDP rather than a POMDP. We have presented the POMDP case to maintain consistency with AIF, but both frameworks can be applied in both MDPs and POMDPs.

second term (state divergence) from the bound. Moreover, under the assumption that the action prior is uniform $p(\mathbf{a}_t|\mathbf{s}_t) := \frac{1}{|A|}$, the action divergence term reduces to the negative entropy of actions. Maximising both reward and an action entropy term provides several benefits, including a mechanism for offline learning, improved exploration and increased algorithmic stability [17,18]. The fourth term (observation ambiguity), which only arises in a POMDP setting, encourages agents to seek out states which have a precise mapping to observations, thus driving agents to regions of observation space where the latent space can be easily inferred, and to avoid regions where the likelihood mapping is highly uncertain.

Traditionally, CAI has been concerned with inferring *policies*, or time-dependent state-action mappings. Here, we reformulate the standard CAI approach to instead infer fixed action *plans* $\pi = \{\mathbf{a}_t, ..., \mathbf{a}_T\}$. Specifically, we derive an N-step planning variational bound for CAI and show that it can be used to derive an expression for the optimal plan. We adapt the generative model and approximate posterior to account for a temporal *sequence* of variables $p(\mathbf{s}_{t:T}, \pi, \mathbf{o}_{t:T}, \mathcal{O}_{t:T}) = \prod_t^T p(\mathcal{O}_t|\mathbf{s}_t, \pi)p(\mathbf{o}_t|\mathbf{s}_t)p(\mathbf{s}_t|\mathbf{s}_{t-1}, \pi)p(\pi)$ and $q(\mathbf{s}_{t:T}, \pi) = \prod_t^T q(\mathbf{s}_t|\pi)q(\pi)$. The optimal plan can then be retrieved as:

$$
\begin{aligned}
\mathcal{L} &= D_{\mathrm{KL}}\Big(q(\mathbf{s}_{t:T}, \pi) \| p(\mathbf{s}_{t:T}, \pi, \mathbf{o}_{t:T}, \mathcal{O}_{t:T})\Big) \\
&= D_{\mathrm{KL}}\Big(q(\pi)\prod_t^T q(\mathbf{s}_t|\pi) \| p(\pi)\prod_t^T p(\mathcal{O}_t|\mathbf{s}_t, \pi)p(\mathbf{o}_t|\mathbf{s}_t)p(\mathbf{s}_t|\mathbf{s}_{t-1}, \pi)\Big) \\
&= D_{\mathrm{KL}}\Big(q(\pi)\sum_t^T D_{\mathrm{KL}}[q(\mathbf{s}_t|\pi) \| p(\mathcal{O}_t|\mathbf{s}_t, \pi)p(\mathbf{o}_t|\mathbf{s}_t)p(\mathbf{s}_t|\mathbf{s}_{t-1}, \pi)] \| p(\pi)\Big) \\
&= D_{\mathrm{KL}}\Big(q(\pi) \| p(\pi)\exp(-\sum_t^T \mathcal{L}_t(\pi))\Big) \implies q^*(\pi) = \sigma\Big(p(\pi) - \sum_t^T \mathcal{L}_t(\pi)\Big)
\end{aligned}
\tag{2}
$$

The optimal plan is thus a path integral of $\mathcal{L}_t(\pi)$, which can be written as:

$$
\begin{aligned}
\mathcal{L}_t(\pi) &= \mathbb{E}_{q(\mathbf{s}_t|\pi)}\big[\ln q(\mathbf{s}_t|\pi) - \ln p(\mathbf{s}_t, \pi, \mathbf{o}_t, \mathcal{O}_t)\big] \\
&= \underbrace{-\mathbb{E}_{q(\mathbf{s}_t|\pi)}\big[\ln p(\mathcal{O}_t|\mathbf{s}_t, \pi)\big]}_{\text{Extrinsic Value}} + \underbrace{D_{\mathrm{KL}}\Big(q(\mathbf{s}_t|\pi) \| p(\mathbf{s}_t|\mathbf{s}_{t-1}, \pi)\Big)}_{\text{State divergence}} - \underbrace{\mathbb{E}_{q(\mathbf{s}_t|\pi)}\big[\ln p(\mathbf{o}_t|\mathbf{s}_t)\big]}_{\text{Observation Ambiguity}}
\end{aligned}
\tag{3}
$$

which is equivalent to Eq. 1 except that it omits the action-divergence term.

4   Active Inference

Unlike CAI, AIF does not introduce additional variables to incorporate 'value' into the generative model. Instead, AIF assumes that the generative model is intrinsically biased towards valuable states or observations. For instance, we might assume that the prior distribution over observations is biased towards observing rewards, $\ln \tilde{p}(\mathbf{o}_{t:T}) \propto e^{r(\mathbf{o}_{t:T})}$, where we use notation $\tilde{p}(\cdot)$ to denote a biased distribution[2]. Let the agent's generative model be defined as $\tilde{p}(\mathbf{s}_{t:T}, \mathbf{o}_{t:T}, \pi) = p(\pi) \prod_t^T p(\mathbf{s}_t | \mathbf{o}_t, \pi) \tilde{p}(\mathbf{o}_t | \pi)$, and the approximate posterior as $q(\mathbf{s}_{t:T}, \pi) = q(\pi) \prod_t^T q(\mathbf{s}_t | \pi)$.

It is then possible to derive an analytical expression for the optimal plan:

$$-\mathcal{F}(\pi) = \mathbb{E}_{q(\mathbf{o}_{t:T}, \mathbf{s}_{t:T}, \pi)} \big[\ln q(\mathbf{s}_{t:T}, \pi) - \ln \tilde{p}(\mathbf{o}_{t:T}, \mathbf{s}_{t:T}, \pi) \big]$$

$$\implies q^*(\pi) = \sigma\big(\ln p(\pi) - \sum_t^T \mathcal{F}_t(\pi) \big) \tag{4}$$

where $-\mathcal{F}_t(\pi)$ is referred to as the *expected free energy* (note that other functionals are also consistent with AIF [23]). Given a uniform prior over policies, behaviour is determined by the expected free energy functional, which decomposes into:

$$-\mathcal{F}_t(\pi) = -\mathbb{E}_{q(\mathbf{o}_t, \mathbf{s}_t | \pi)} \big[\ln q(\mathbf{s}_t | \pi) - \ln \tilde{p}(\mathbf{o}_t, \mathbf{s}_t | \pi) \big]$$

$$= \underbrace{-\mathbb{E}_{q(\mathbf{o}_t, \mathbf{s}_t | \pi)} \big[\ln \tilde{p}(\mathbf{o}_t | \pi) \big]}_{\text{Extrinsic Value}} - \underbrace{\mathbb{E}_{q(\mathbf{o}_t | \pi)} \big[D_{\mathrm{KL}}\big(q(\mathbf{s}_t | \mathbf{o}_t, \pi) \| q(\mathbf{s}_t | \pi) \big) \big]}_{\text{Intrinsic Value}} \tag{5}$$

where we have made the assumption that the inference procedure is approximately correct, such that $q(\mathbf{s}_t | \mathbf{o}_t, \pi) \approx p(\mathbf{s}_t | \mathbf{o}_t, \pi)$. This assumption is not unreasonable since it merely presupposes that action selection occurs after perceptual inference, which directly attempts to minimize the divergence $D_{\mathrm{KL}}\big(q(\mathbf{s}_t | \mathbf{o}_t, \pi) \| p(\mathbf{s}_t | \mathbf{o}_t, \pi) \big)$ between the approximate and true posterior. Because agents are required to minimise Eq. 5, they are required to maximise both extrinsic and intrinsic value. Extrinsic value measures the degree to which expected observations are consistent with prior beliefs about favourable observations. Under the assumption that $\ln \tilde{p}(\mathbf{o}_{t:T}) \propto e^{r(\mathbf{o}_{t:T})}$, this is equivalent to seeking out rewarding observations. Intrinsic value is equivalent to the expected information gain over states, which compels agents to seek informative observations which maximally reduce uncertainty about hidden states.

While AIF is usually formulated in terms of fixed action sequences (i.e. plans), it can also be formulated in terms of policies (i.e. state-action mappings). Let the agent's generative model be defined as $\tilde{p}(\mathbf{s}_t, \mathbf{o}_t, \mathbf{a}_t) = p(\mathbf{s}_t | \mathbf{o}_t, \mathbf{a}_t) p(\mathbf{a}_t | \mathbf{s}_t) \tilde{p}(\mathbf{o}_t | \mathbf{a}_t)$, and the approximate posterior as $q_\phi(\mathbf{s}_t, \mathbf{a}_t) = q_\phi(\mathbf{a}_t | \mathbf{s}_t) q(\mathbf{s}_t)$. We can now write the expected free energy functional in terms of the policy parameters ϕ:

---

[2] AIF is usually formulated solely in terms of observations, such that some observations are more 'favourable' than others. We introduced the notion of rewards to retain consistency with CAI.

$$-\mathcal{F}_t(\phi) = \mathbb{E}_{q(\mathbf{o}_t,\mathbf{s}_t,\mathbf{a}_t)}\Big[\ln q_\phi(\mathbf{a}_t,\mathbf{s}_t) - \ln \tilde{p}(\mathbf{s}_t,\mathbf{o}_t,\mathbf{a}_t)\Big]$$

$$= -\underbrace{\mathbb{E}_{q(\mathbf{o}_t|\mathbf{a}_t)}\Big[\ln \tilde{p}(\mathbf{o}_t|\mathbf{a}_t)\Big]}_{\text{Extrinsic Value}} - \underbrace{\mathbb{E}_{q(\mathbf{o}_t,\mathbf{a}_t|\mathbf{s}_t)}\Big[D_{\text{KL}}\big(q(\mathbf{s}_t|\mathbf{o}_t,\mathbf{a}_t)\|q(\mathbf{s}_t|\mathbf{a}_t)\big)\Big]}_{\text{Intrinsic Value}} \quad (6)$$

$$+ \underbrace{\mathbb{E}_{q(\mathbf{s}_t)}\Big[D_{\text{KL}}\big(q_\phi(\mathbf{a}_t|\mathbf{s}_t)\|p(\mathbf{a}_t|\mathbf{s}_t)\big)\Big]}_{\text{Action Divergence}}$$

Inferring policies with AIF thus requires minimizing an action divergence term.

5   Encoding Value

The previous sections demonstrate that both AIF and CAI can be formulated as variational inference, for both fixed action sequences (i.e. plans) and policies (i.e. state-action mappings). We now move on to consider the key difference between these frameworks – how they encode 'value'. AIF encodes value directly into the generative model as a prior over observations, whereas in CAI, extrinsic value is effectively encoded into the likelihood which, by Bayes rule, relates to the prior as $p(\mathbf{o}|\mathbf{s}) = p(\mathbf{o})\frac{p(\mathbf{s})}{p(\mathbf{s}|\mathbf{o})}$. When applied within a KL divergence, this fraction becomes a negative information gain. We elucidate this distinction by introducing a further variant of active inference, which here we call *likelihood-AIF*, where instead of a biased prior over rewards the agent has a biased likelihood $\tilde{p}(\mathbf{o}_t,\mathbf{s}_t) = \tilde{p}(\mathbf{o}_t|\mathbf{s}_t)p(\mathbf{s}_t)$. The likelihood-AIF objective functional $\hat{\mathcal{F}}(\phi)$ becomes:

$$-\hat{\mathcal{F}}_t(\phi) = \mathbb{E}_{q_\phi(\mathbf{s}_t,\mathbf{o}_t,\mathbf{a}_t)}\Big[\ln q_\phi(\mathbf{s}_t,\mathbf{a}_t) - \ln \tilde{p}(\mathbf{o}_t,\mathbf{s}_t,\mathbf{a}_t)\Big]$$

$$= -\underbrace{\mathbb{E}_{q_\phi(\mathbf{s}_t,\mathbf{a}_t)}\Big[\ln \tilde{p}(\mathbf{o}_t|\mathbf{s}_t)\Big]}_{\text{Extrinsic Value}} + \underbrace{D_{\text{KL}}\Big(q(\mathbf{s}_t)\|p(\mathbf{s}_t|\mathbf{s}_{t-1},\mathbf{a}_{t-1})\Big)}_{\text{State divergence}} + \underbrace{D_{\text{KL}}\Big(q_\phi(\mathbf{a}_t|\mathbf{s}_t)\|p(\mathbf{a}_t|\mathbf{s}_t)\Big)}_{\text{Action Divergence}}$$

If we set $\ln \tilde{p}(\mathbf{o}_t|\mathbf{s}_t) = \ln p(\mathcal{O}_t|\mathbf{s}_t,\mathbf{a}_t)$, this is exactly equivalent to the CAI objective in the case of MDPs. The fact that likelihood AIF on POMDPs is equivalent to CAI on MDPs is due to the fact that the observation modality in AIF is 'hijacked' by the encoding of value, and thus effectively contains one less degree-of-freedom compared to CAI, which maintains a separate veridical representation of observation likelihoods. A further connection is that AIF on MDPs is equivalent to KL control [8,26–28], and the recently proposed state-marginal-matching [19] objectives. We leave further exploration of these similarities to future work.

6   Discussion

In this work, we have highlighted the large degree of overlap between the frameworks of active inference (AIF) and control as inference (CAI), and we have explored the major way in which they differ - which is in terms of how they encode value into their generative models, thus turning a value-free inference problem into one that can serve the purposes of adaptive action. While CAI augments the 'natural' probabilistic graphical model with exogenous optimality variables.[3], AIF leaves the structure of the graphical model unaltered and instead encodes value into the generative model directly. These two approaches lead to significant differences between their respective functionals. AIF, by contaminating its generative model with value-imbuing biases, loses a degree of freedom compared to CAI, which maintains a strict separation between an ideally veridical generative model of the environment and the goals of the agent. In POMDPs, this approach results in CAI incorporating an 'observation-ambiguity' term which is absent in the AIF formulation. Secondly, the different methods for encoding the probability of goals – likelihoods in CAI and priors in AIF – lead to different exploratory terms in the objective functionals. Specifically, AIF is endowed with an expected information gain that CAI lacks. AIF approaches thus lend themselves naturally to goal-directed exploration whereas CAI induces only random, entropy-maximizing exploration. Moreover, when AIF is applied to infer actions directly, it also obtains the same action-entropy terms as CAI, while additionally requiring AIF agents to maximize exploratory terms. When CAI is extended to the POMDP setting, it gives rise to an additional observation-ambiguity term to be minimized, which drives agent to seek out states with highly precise likelihood mappings, which in effect penalizes exploration. Thus, AIF encourages exploration while maintaining a biased perceptual system, while CAI explores randomly but maintains a principled separation between veridical perception and control.

These different ways of encoding goals into probabilistic models also lend themselves to more philosophical interpretations. By viewing goals as an additional exogenous factor in an otherwise unbiased inference process, CAI maintains the modularity thesis of separate perception and action modules [3]. This makes CAI approaches deeply consonant with the mainstream view in machine learning that sees the goal of perception as recovering veridical representations of the world, and control as using this world-model to plan actions. In contrast, AIF elides these clean boundaries between unbiased perception and action by instead positing that *biased* perception is crucial to adaptive action. Rather than maintaining an unbiased world model that predicts likely consequences, AIF instead maintains a biased generative model which preferentially predicts the agent's preferences being fulfilled. Active Inference thus aligns closely with enactive and embodied approaches [4, 9] to cognition, which view the action-perception loop as a continual flow rather than a sequence of distinct stages.

We have thus seen how two means of encoding preferences into inference problems leads to two distinct families of algorithms, each optimising subtly different functionals, resulting in differing behaviour. This raises the natural questions of which method should be preferred, and whether these are the only two

---

[3] Utilising optimality variables is not strictly necessary for CAI. In the case of undirected graphical models, an additional undirected factor can be appended to each node [38]. Interestingly, this approach bears similarities to the procedure adopted in [24], suggesting a further connection between generalised free energy and CAI.

possible methods. One can imagine explicitly modelling the expected reward, and biasing inferences with priors over the reward. Alternatively, agents could maintain desired distributions over states, observations, and actions, which would maximize the flexibility in specifying goals intrinsic to the variational control framework. These potential extensions to the framework, their relation to one another, and the objective functionals they induce, are topics for future work.

References

1. Abdolmaleki, A., Springenberg, J.T., Tassa, Y., Munos, R., Heess, N., Riedmiller, M.: Maximum a posteriori policy optimisation. arXiv preprint arXiv:1806.06920 (2018)
2. Attias, H.: Planning by probabilistic inference. In: AISTATS. Citeseer (2003)
3. Baltieri, M., Buckley, C.L.: The modularity of action and perception revisited using control theory and active inference. In: Artificial Life Conference Proceedings, pp. 121–128. MIT Press (2018)
4. Baltieri, M., Buckley, C.L.: Generative models as parsimonious descriptions of sensorimotor loops. arXiv preprint arXiv:1904.12937 (2019)
5. Beal, M.J.: Variational algorithms for approximate Bayesian inference. Ph.D. thesis, UCL (University College London) (2003)
6. Blackmore, P.A., Bitmead, R.R.: Duality between the discrete-time Kalman filter and LQ control law. IEEE Trans. Autom. Control $40(8)$, 1442–1444 (1995)
7. Botvinick, M., Toussaint, M.: Planning as inference. Trends Cogn. Sci. $16(10)$, 485–488 (2012)
8. van den Broek, L., Wiegerinck, W., Kappen, H.J.: Risk sensitive path integral control (2010)
9. Clark, A.: Radical predictive processing. South. J. Philos. 53, 3–27 (2015)
10. Fountas, Z., Sajid, N., Mediano, P.A., Friston, K.: Deep active inference agents using Monte-Carlo methods. arXiv preprint arXiv:2006.04176 (2020)
11. Friston, K.: The free-energy principle: a unified brain theory? Nat. Rev. Neurosci. $11(2)$, 127–138 (2010)
12. Friston, K., FitzGerald, T., Rigoli, F., Schwartenbeck, P., Pezzulo, G.: Active inference: a process theory. Neural Comput. $29(1)$, 1–49 (2017)
13. Friston, K., Kilner, J., Harrison, L.: A free energy principle for the brain. J. Physiol. Paris $100(1$–$3)$, 70–87 (2006)
14. Friston, K., Rigoli, F., Ognibene, D., Mathys, C., Fitzgerald, T., Pezzulo, G.: Active inference and epistemic value. Cogn. Neurosci. $6(4)$, 187–214 (2015)
15. Friston, K.J., Daunizeau, J., Kiebel, S.J.: Reinforcement learning or active inference? PLoS ONE $4(7)$ (2009)
16. Friston, K.J., Parr, T., de Vries, B.: The graphical brain: belief propagation and active inference. Netw. Neurosci. $1(4)$, 381–414 (2017)
17. Haarnoja, T., Zhou, A., Abbeel, P., Levine, S.: Soft actor-critic: off-policy maximum entropy deep reinforcement learning with a stochastic actor. arXiv preprint arXiv:1801.01290 (2018)
18. Haarnoja, T., et al.: Soft actor-critic algorithms and applications. arXiv preprint arXiv:1812.05905 (2018)
19. Lee, L., Eysenbach, B., Parisotto, E., Xing, E., Levine, S.: Efficient exploration via state marginal matching. arXiv preprint arXiv:1906.05274 (2019)

20. Levine, S.: Reinforcement learning and control as probabilistic inference: tutorial and review. arXiv preprint arXiv:1805.00909 (2018)
21. Millidge, B.: Combining active inference and hierarchical predictive coding: a tutorial introduction and case study (2019)
22. Millidge, B.: Implementing predictive processing and active inference: preliminary steps and results (2019)
23. Millidge, B., Tschantz, A., Buckley, C.L.: Whence the expected free energy? arXiv preprint arXiv:2004.08128 (2020)
24. Parr, T., Friston, K.J.: Generalised free energy and active inference. Biol. Cybern. **113**(5–6), 495–513 (2019)
25. Pio-Lopez, L., Nizard, A., Friston, K., Pezzulo, G.: Active inference and robot control: a case study. J. R. Soc. Interface **13**(122), 20160616 (2016)
26. Rawlik, K., Toussaint, M., Vijayakumar, S.: Approximate inference and stochastic optimal control. arXiv preprint arXiv:1009.3958 (2010)
27. Rawlik, K., Toussaint, M., Vijayakumar, S.: On stochastic optimal control and reinforcement learning by approximate inference. In: Twenty-Third International Joint Conference on Artificial Intelligence (2013)
28. Rawlik, K.C.: On probabilistic inference approaches to stochastic optimal control (2013)
29. Sancaktar, C., Lanillos, P.: End-to-end pixel-based deep active inference for body perception and action. arXiv preprint arXiv:2001.05847 (2019)
30. Theodorou, E.A., Todorov, E.: Relative entropy and free energy dualities: connections to path integral and KL control. In: 2012 IEEE 51st IEEE Conference on Decision and Control (CDC), pp. 1466–1473. IEEE (2012)
31. Todorov, E.: General duality between optimal control and estimation. In: 2008 47th IEEE Conference on Decision and Control, pp. 4286–4292. IEEE (2008)
32. Tschantz, A., Baltieri, M., Seth, A., Buckley, C.L., et al.: Scaling active inference. arXiv preprint arXiv:1911.10601 (2019)
33. Tschantz, A., Millidge, B., Seth, A.K., Buckley, C.L.: Reinforcement learning through active inference. arXiv preprint arXiv:2002.12636 (2020)
34. Tschantz, A., Seth, A.K., Buckley, C.L.: Learning action-oriented models through active inference. PLoS Comput. Biol. **16**(4), e1007805 (2020)
35. Ueltzhöffer, K.: Deep active inference. Biol. Cybern. **112**(6), 547–573 (2018)
36. Wainwright, M.J., Jordan, M.I.: Graphical models, exponential families, and variational inference. Now Publishers Inc. (2008)
37. Walsh, K.S., McGovern, D.P., Clark, A., O'Connell, R.G.: Evaluating the neurophysiological evidence for predictive processing as a model of perception. Ann. N. Y. Acad. Sci. **1464**(1), 242 (2020)
38. Ziebart, B.D.: Modeling purposeful adaptive behavior with the principle of maximum causal entropy (2010)

Active Inference or Control as Inference?
A Unifying View

Abraham Imohiosen[1], Joe Watson[2(✉)], and Jan Peters[2]

[1] RWTH Aachen University, Aachen, Germany
abraham.imohiosen@rwth-aachen.de
[2] IAS, Technical University Darmstadt, Darmstadt, Germany
{watson,peters}@ias.informatik.tu-darmstadt.de

Abstract. Active inference (AI) is a persuasive theoretical framework from computational neuroscience that seeks to describe action and perception as inference-based computation. However, this framework has yet to provide practical sensorimotor control algorithms that are competitive with alternative approaches. In this work, we frame active inference through the lens of control as inference (CaI), a body of work that presents trajectory optimization as inference. From the wider view of 'probabilistic numerics', CaI offers principled, numerically robust optimal control solvers that provide uncertainty quantification, and can scale to nonlinear problems with approximate inference. We show that AI may be framed as partially-observed CaI when the cost function is defined specifically in the observation states.

Keywords: Active inference · Control · Approximate inference

1 Introduction

Active inference (AI) [2,4,5] is a probabilistic framework for sensorimotor behavior that enjoyed sustained interest from computational neuroscientists. However, its formulation has been criticized for its opacity and similarity to optimal control [7–9], but is seemingly difficult to translate into an equally effective algorithmic form. In this work, we offer a critical analysis of AI from the view of control as inference (CaI) [1,11,14,21,24,28], the synthesis of optimal control and approximate inference. The goal is to appreciate the insights from the AI literature, but in a form with computational and theoretical clarity.

2 Background

Here we outline the foundational theory and assumptions in this work.

© Springer Nature Switzerland AG 2020
T. Verbelen et al. (Eds.): IWAI 2020, CCIS 1326, pp. 12–19, 2020.
https://doi.org/10.1007/978-3-030-64919-7_2

2.1   Problem Formulation

We specifically consider a known stochastic, continuous, discrete-time, partially-observed, nonlinear, dynamical system with state $x \in \mathbb{R}^{d_x}$, observations $y \in \mathbb{R}^{d_y}$ and control inputs $u \in \mathbb{R}^{d_u}$, operating over a time horizon T. We define the states in upper case to denote the variables over the time horizon, i.e. $U = \{u_0, \ldots, u_{T-1}\}$. The joint distribution (generative model) $p(Y, X, U)$ over these variables factorizes into several interpretable distributions: The dynamics $p(x_{t+1}|x_t, u_t)$, observation model $p(y_t \mid x_t, u_t)$, and behavior policy $p(u_t \mid x_t)$.

2.2   Variational Inference for Latent Variable Models

Inference may be described by minimizing the distance between the 'true' data distribution $p(\cdot)$ and a parameterized family $q_\theta(\cdot)$ [17]. A popular approach is to minimize the Kullback-Liebler (KL) divergence, e.g. $\min D_{\mathrm{KL}}[q_\theta \parallel p]$ w.r.t. θ. More complex inference tasks can be described by observations y influenced by unseen latent variables x. Given an observation y^*, maximizing the likelihood involves integrating over the hidden states, and so is termed the marginal likelihood $p(y^*) = \int p(y = y^*, x) dx$. Unfortunately this marginalization is typically intractable in closed-form. A more useful objective may be obtained by applying a variational approximation of latent state $q_\theta(x \mid y^*) = q_\theta(x \mid y = y^*)$ to the log marginal likelihood and obtaining a lower bound via Jensen's inequality [17]

$$\log \int p(y^*, x) dx = \log \int p(y^*, x) \tfrac{q_\theta(x|y^*)}{q_\theta(x|y^*)} dx = \log \mathbb{E}_{x \sim q_\theta(\cdot|y^*)} \left[\tfrac{p(y^*, x)}{q_\theta(x|y^*)} \right], \quad (1)$$

$$\geq \mathbb{E}_{x \sim q_\theta(\cdot|y^*)} \left[\log \tfrac{p(y^*, x)}{q_\theta(x|y^*)} \right] = -D_{\mathrm{KL}}[q_\theta(x \mid y^*) \| p(x, y^*))], \quad (2)$$

$$= \mathbb{E}_{x \sim q_\theta(\cdot|y^*)}[\log p(y^* \mid x)] - D_{\mathrm{KL}}[q_\theta(x \mid y^*) \| p(x)], \quad (3)$$

where Eqs. 2, 3 are variations of the 'evidence lower bound objective' (ELBO). The expectation maximization algorithm (EM) [17], can be understood via Eq. 3 as iteratively estimating the latent states (minimizing the KL term via q) in the E step and maximizing the likelihood term in the M step.

3   Active Inference

Active Inference frames sensorimotor behaviour as the goal of equilibrium between its current and desired observations, which in practice can be expressed as the minimization of a distance between these two quantities. This distance is expressed using the KL divergence, resulting in a variational free energy objective as described in Sect. 2.2. Curiously, AI is motivated directly by the ELBO, whose negative is referred to in the AI literature as the 'free energy' $\mathcal{F}(\cdot)$. The minimization of this quantity, $\mathcal{F}(y^*, x, u) = D_{\mathrm{KL}}[q_\theta(x, u \mid y^*) \| p(x, u, y^*)]$, as a model of behavior (i.e. state estimation and control), has been coined the 'free energy principle'.

3.1 Free Energy of the Future

Despite the ELBO not being temporally restricted, AI delineates a 'future' free energy. This free energy is used to describe the distance between future predicted and desired observations, where u is directly represented as a policy $u = \pi(x)$, so $\mathcal{F}(y_t^*, x_t \mid \pi)$ over the future trajectory is minimized. In active inference, π is commonly restricted to discrete actions or an ensemble of fixed policies, so inferring $p(\pi)$ can be approximated through a softmax $\sigma(\cdot)$ applied to the expected 'future' free energies for each policy over $t = [\tau, \ldots, T-1]$, with temperature γ and prior $p(\pi)$

$$p(\pi \mid \boldsymbol{Y}^*) \approx \sigma(\log p(\pi) + \gamma \textstyle\sum_{t=\tau}^{T-1} \mathcal{F}(\boldsymbol{y}_t^*, \boldsymbol{x}_t, \mid \pi)). \tag{4}$$

Moreover, for the 'past' where $t = [0, \ldots, \tau - 1]$, minimizing $\mathcal{F}(\cdot)$ amounts for state estimation of x given y. Another consideration is whether the dynamic and observation models are known or unknown. In this work we assume they are given, but AI can also include estimating these models from data.

3.2 Active Inference in Practice

Initial AI work was restricted to discrete domains and evaluated on simple grid-world environments [5,6]. Later work on continuous state spaces use various black-box approaches such as cross-entropy [25], evolutionary strategies [26], and policy gradient [16] to infer π. A model-based method was achieved by using stochastic VI on expert data [3]. Connections between AI and CaI, performing inference via message passing, have been previously discussed [13,27]. AI has been applied to real robots for kinematic planning, performing gradient descent on the free energy using the Laplace approximation every timestep [18]. Despite these various approaches, AI has yet to demonstrate the sophisticated control achieved by advanced optimal methods, such as differential dynamic programming [20].

4 Control as Inference

From its origins in probabilistic control design [12], defining a state $z \in \mathbb{R}^{d_z}$ to describe the desired system trajectory[1] $p(\boldsymbol{Z})$, optimal control can be expressed as finding the state-action distribution that minimizes the distance for a generative model parameterized by $\boldsymbol{\theta}$, which can be framed as a likelihood objective [17]

$$\min D_{\mathrm{KL}}[p(\boldsymbol{Z}) \parallel q_\theta(\boldsymbol{Z})] \equiv \max \ \mathbb{E}_{\boldsymbol{Z} \sim p(\cdot)}[\log \textstyle\int q_\theta(\boldsymbol{Z}, \boldsymbol{X}, \boldsymbol{U}) d\boldsymbol{X} d\boldsymbol{U}]. \tag{5}$$

When $p(\boldsymbol{Z})$ simply describes a desired state $\boldsymbol{z}_t^*$, so $p(\boldsymbol{z}_t) = \delta(\boldsymbol{z}_t - \boldsymbol{z}_t^*)$, and the latent state-action trajectory is approximated by $q_\phi(\boldsymbol{X}, \boldsymbol{U})$, the objective (Eq. 5) can be expressed as an ELBO where the 'data' is $\boldsymbol{Z}^*$

$$\max \mathbb{E}_{\boldsymbol{X}, \boldsymbol{U} \sim q_\phi(\cdot \mid \boldsymbol{Z}^*)}[\log q_\theta(\boldsymbol{Z}^* \mid \boldsymbol{X}, \boldsymbol{U})] - D_{\mathrm{KL}}[q_\phi(\boldsymbol{X}, \boldsymbol{U} \mid \boldsymbol{Z}^*) \mid q_\theta(\boldsymbol{X}, \boldsymbol{U})], \tag{6}$$

---

[1] While z could be defined from $[x, u]^\mathsf{T}$, it could also include a transformation, e.g. applying kinematics to joint space-based control for a cartesian space objective.

where ϕ captures the latent state parameterization and θ defines the remaining terms, i.e. the priors on the system parameters and latent states. This objective can be optimized using EM, estimating the latent state-action trajectory ϕ in the E step and optimizing the remaining unknowns θ in the M step. By exploiting the temporal structure, $q_\phi(X, U \mid Z^*)$ can be inferred efficiently in the E step by factorizing the joint distribution (Eq. 7) and applying Bayes rule recursively

$$q_\phi(Z^*, X, U) = q_\phi(x_0) \prod_{t=0}^{T-1} q_\phi(x_{t+1}|x_t, u_t) \prod_{t=0}^{T} q_\phi(z_t^*|x_t, u_t) q_\phi(u_t|x_t), \tag{7}$$

$$q_\phi(x_t, u_t \mid z_{0:t}^*) \propto q_\phi(z_t^* \mid x_t, u_t) \, q_\phi(x_t, u_t \mid z_{0:t-1}^*), \tag{8}$$

$$q_\phi(x_t, u_t \mid z_{0:T}^*) \propto q_\phi(x_t, u_t \mid x_{t+1}) \, q_\phi(x_{t+1} \mid z_{0:T}^*). \tag{9}$$

Eqs. 8, 9 are commonly known as Bayesian filtering and smoothing [19]. The key distinction of this framework from state estimation is the handling of u during the forward pass, as $q_\phi(x_t, u_t) = q_\phi(u_t \mid x_t) q_\phi(x_t)$, control is incorporated into the inference. We can demonstrate this in closed-form with linear Gaussian inference and linear quadratic optimal control.

4.1 Linear Gaussian Inference and Linear Quadratic Control

While the formulation above is intentionally abstract, it can be grounded clearly by unifying linear Gaussian dynamical system inference (LGDS, i.e. Kalman filtering and smoothing) and linear quadratic Gaussian (LQG) optimal control [22]. While both cases have linear dynamical systems, here LQG is fully-observed[2] and has a quadratic control cost, while the LGDS is partially observed and has a quadratic log-likelihood due to the Gaussian additive uncertainties. These two domains can be unified by viewing the quadratic control cost function as an Gaussian observation likelihood. For example, given $z_t = x_t + \xi, \xi \sim \mathcal{N}(0, \Sigma)$ and $z_t^* = 0 \; \forall \, t$,

$$\log q_\theta(z_t^*|x_t, u_t) = -\tfrac{1}{2}(d_z \log 2\pi + \log|\Sigma| + x_t^\mathsf{T} \Sigma^{-1} x_t) = \alpha x_t^\mathsf{T} Q x_t + \beta \tag{10}$$

where (α, β) represents the affine transformation mapping the quadratic control cost $x^\mathsf{T} Q x$ to the Gaussian likelihood. As convex objectives are invariant to affine transforms, this mapping preserves the control problem while translating it into an inference one. The key unknown here is α, which incorporates Q into the additive uncertainty ξ, $\Sigma = \alpha Q^{-1}$. Moreover, inference is performed by using message passing [15] in the E step to estimate X and U, while α is optimized in the M step. This view scales naturally to not just the typical LQG

---

[2] Confusingly, LQG can refer to both Gaussian disturbance and/or observation noise. While all varieties share the same optimal solution as LQR, the observation noise case results in a partially observed system and therefore requires state estimation. I2C is motivated by the LQR solution and therefore does not consider observation noise, but it would be straightforward to integrate.

cost $x^\mathsf{T}Qx + u^\mathsf{T}Ru$, but also nonlinear mappings to z by using approximate inference. While the classic LQG result includes the backward Ricatti equations and an optimal linear control law, the inference setting derives direct parallels to the backward pass during smoothing [22] and the linear conditional distribution of the Gaussian, $q_\theta(u_t \mid x_t) = \mathcal{N}(K_t x_t + k_t, \Sigma_{k_t})$ [10] respectively. As the conditional distribution is linear, updating the prior joint density $p(x_t, u_t)$ in the forward pass with updated state estimate x_t' corresponds to linear feedback control w.r.t. the prior

$$p(u_t') = \int p(u_t|x_t = x_t')p(x_t')dx_t', \qquad (11)$$

$$\mu_{u_t'} = \mu_{u_t} + K_t(\mu_{x_t} - \mu_{x_t'}), \qquad (12)$$

$$\Sigma_{uu_t'} = \Sigma_{uu_t} - \Sigma_{ux_t}\Sigma_{xx_t}^{-1}\Sigma_{xu_t}^\mathsf{T} + K_t\Sigma_{xx_t'}K_t^\mathsf{T}, \qquad (13)$$

$$K_t = \Sigma_{ux_t}\Sigma_{xx_t}^{-1}. \qquad (14)$$

From Eq. 14, it is evident that the strength of the feedback control depends on both the certainty in the state and the correlation between the optimal state and action.

The general EM algorithm for obtaining $q_\theta(x, u)$ from $p(Z)$ is referred to as input inference for control (I2C) [28] due to its equivalence with input estimation. Note that for linear Gaussian EM, the ELBO is tight as the variational distribution is the exact posterior. For nonlinear filtering and smoothing, mature approximate inference methods such as Taylor approximations, quadrature and sequential Monte Carlo may be used for efficient and accurate computation [19].

Another aspect to draw attention to is the inclusion of z compared to alternative CaI formulations, which frame optimality as the probability for some discrete variable o, $p(o = 1 \mid x, u)$ [14]. Previous discussion on CaI vs AI have framed this discrete variable as an important distinction. However, it is merely a generalization to allow for a general cost function $C(\cdot)$ to be framed as a log-likelihood, i.e. $p(o = 1 \mid x, u) \propto \exp(-\alpha C(x, u))$. For the typical state-action cost functions that are a distance metric in some transformed space, the key consideration is the choice of observation space z and corresponding exponential density.

5   The Unifying View: Control of the Observations

A key distinction to the AI and CaI formulations described above is that, while AI combines state estimation and control with a unified objective, CaI focuses on trajectory optimization. However, this need not be the case. In a similar fashion to the partially-observed case of LQG, CaI also naturally incorporates observations [23]. As Sect. 4 describes I2C through a general Bayesian dynamical system, the formulation can be readily adapted to include inference using past measurements. Moreover, as I2C frames the control objective as an observation likelihood, when z and y are the same transform of x and u, the objective can also be unified and directly compared to active inference. For 'measurements' $\boldsymbol{Y}^* = \{\boldsymbol{y}_0^*, \ldots, \boldsymbol{y}_{\tau-1}^*, \boldsymbol{z}_\tau^*, \ldots, \boldsymbol{z}_{T-1}^*\}$, following Eq. 5 using the $\mathcal{F}(\cdot)$ notation

$$\min D_{\mathrm{KL}}[p(\boldsymbol{Y}) \| q_\theta(\boldsymbol{Y})] = \min \underbrace{\sum_{t=0}^{\tau-1} \mathcal{F}_\psi(\boldsymbol{y}_t^*, \boldsymbol{x}_t, \boldsymbol{u}_t)}_{\text{state estimation}} + \underbrace{\sum_{t=\tau}^{T-1} \mathcal{F}_\psi(\boldsymbol{z}_t^*, \boldsymbol{x}_t, \boldsymbol{u}_t)}_{\text{optimal control}},$$

(15)

where $\psi = \{\theta, \phi\}$. Here, $p(\boldsymbol{y}_t) = \delta(\boldsymbol{y}_t - \boldsymbol{y}_t^*)$ now also describes the empirical density of past measurements $\boldsymbol{y}_{<\tau}^*$. The crucial detail for this representation is that the observation model $q_\theta(\boldsymbol{y}_t \mid \boldsymbol{x}_t, \boldsymbol{u}_t, t)$ is now time dependent, switching from estimation to control at $t = \tau$. For the Gaussian example in Sect. 4.1, $\boldsymbol{\Sigma}_{<\tau}$ is the measurement noise and $\boldsymbol{\Sigma}_{\geq\tau}^{-1} = \alpha\boldsymbol{Q}$. A benefit of this view is that the computation of active inference can now be easily compared to the classic results of Kalman filtering and LQG (Fig. 1), and also scaled to nonlinear tasks through approximate inference. Moreover, obtaining the policy $\pi(\cdot)$ using the joint distribution $q_\theta(\boldsymbol{x}_t, \boldsymbol{u}_t)$ is arguably a more informed approach compared to direct policy search on an arbitrary policy class.

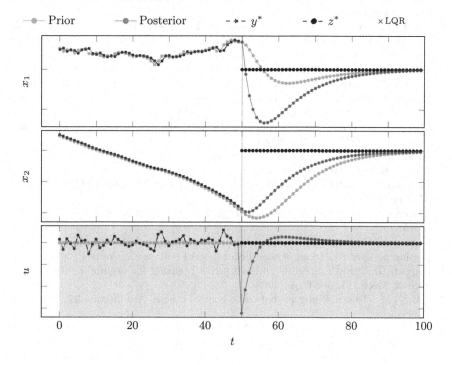

Fig. 1. Linear Gaussian i2c performing state estimation and control following Sect. 5, with state $\boldsymbol{x} = [x_1, x_2]^\mathsf{T}$, action u and $[\boldsymbol{x}, u]^\mathsf{T}$ as the observation space. With $\tau = 50$, for $t < \tau$ i2c performs state estimation under random controls. For $t \geq \tau$, i2c switches to optimal control. This example is in the low noise setting, with a large prior on u, to illustrate that i2c returns the LQR solution for the same initial state and planning horizon.

6 Conclusion

We have derived an equivalent formulation to active inference by considering partially-observed, inference-based optimal control, which has a principled derivation and is well-suited for approximate inference. While we have delineated state estimation as operating on past measurement and control as planning future actions (Eq. 15), both AI and I2C demonstrate the duality between estimation and control due to the mathematical similarity when both are treated probabilistically. We hope the inclusion of the CaI literature enables a greater theoretical understanding of AI and more effective implementations through approximate inference.

References

1. Attias, H.: Planning by probabilistic inference. In: International Workshop on Artificial Intelligence and Statistics (2003)
2. Biehl, M., Guckelsberger, C., Salge, C., Smith, S.C., Polani, D.: Expanding the active inference landscape: more intrinsic motivations in the perception-action loop. Front. Neurorobotics (2018)
3. Catal, O., Nauta, J., Verbelen, T., Simoens, P., Dhoedt, B.: Bayesian policy selection using active inference. In: ICLR Workshop on Structure & Priors in Reinforcement Learning (2019)
4. Friston, K.: The free-energy principle: a unified brain theory? Nat. Rev. Neurosci. **11**, 127–138 (2010)
5. Friston, K., FitzGerald, T., Rigoli, F., Schwartenbeck, P., Pezzulo, G.: Active inference: a process theory. Neural Comput. **29**, 1–49 (2017)
6. Friston, K.J., Daunizeau, J., Kiebel, S.J.: Reinforcement learning or active inference? PLoS ONE **4**, e6421 (2009)
7. Gershman, S.J.: What does the free energy principle tell us about the brain? Neurons, Behavior, Data analysis, and Theory (2019)
8. Guzmán, N.: twitter.com/NoahGuzman14/status/1259953086241492992. Accessed 17 June 2020
9. Herreros, I., Verschure, P.F.: About the goal of a goals' goal theory. Cogn. Neurosci. (2015)
10. Hoffmann, C., Rostalski, P.: Linear optimal control on factor graphs - a message passing perspective. International Federation of Automatic Control (2017)
11. Kappen, H.J.: Path integrals and symmetry breaking for optimal control theory. J. Stat. Mech. Theory Exp. (2005)
12. Kárný, M.: Towards fully probabilistic control design. Automatica **32**, 1719–1722 (1996)
13. van de Laar, T., Özçelikkale, A., Wymeersch, H.: Application of the free energy principle to estimation and control. arXiv preprint arXiv:1910.09823 (2019)
14. Levine, S.: Reinforcement learning and control as probabilistic inference: tutorial and review. arXiv preprint arXiv:1805.00909 (2018)
15. Loeliger, H.A., Dauwels, J., Hu, J., Korl, S., Ping, L., Kschischang, F.R.: The factor graph approach to model-based signal processing. Proc. IEEE **95**, 1295–1322 (2007)
16. Millidge, B.: Deep active inference as variational policy gradients. arXiv preprint arXiv:1907.03876 (2019)

17. Murphy, K.P.: Machine Learning: A Probabilistic Perspective. MIT Press, Cambridge (2012)
18. Oliver, G., Lanillos, P., Cheng, G.: Active inference body perception and action for humanoid robots. arXiv preprint arXiv:1906.03022 (2019)
19. Srkk, S.: Bayesian Filtering and Smoothing. Cambridge University Press, Cambridge (2013)
20. Tassa, Y., Erez, T., Todorov, E.: Synthesis and stabilization of complex behaviors through online trajectory optimization. In: International Conference on Intelligent Robots and Systems. IEEE (2012)
21. Todorov, E.: Linearly-solvable Markov decision problems. In: Advances in Neural Information Processing Systems (2007)
22. Toussaint, M.: Robot trajectory optimization using approximate inference. In: International Conference on Machine Learning (2009)
23. Toussaint, M., Charlin, L., Poupart, P.: Hierarchical POMDP controller optimization by likelihood maximization. In: Uncertainty in Artificial Intelligence (2008)
24. Toussaint, M., Storkey, A.: Probabilistic inference for solving discrete and continuous state Markov decision processes. In: International Conference on Machine Learning (2006)
25. Tschantz, A., Millidge, B., Seth, A.K., Buckley, C.L.: Reinforcement learning through active inference. arXiv preprint arXiv:2002.12636 (2020)
26. Ueltzhöffer, K.: Deep active inference. Biol. Cybern. 112(6), 547–573 (2018). https://doi.org/10.1007/s00422-018-0785-7
27. de Vries, B., Friston, K.J.: A factor graph description of deep temporal active inference. Front. Comput. Neurosci. 11, 95 (2017)
28. Watson, J., Abdulsamad, H., Peters, J.: Stochastic optimal control as approximate input inference. In: Conference on Robot Learning (2019)

Active Inference for Fault Tolerant Control of Robot Manipulators with Sensory Faults

Corrado Pezzato[1(✉)], Mohamed Baioumy[3], Carlos Hernández Corbato[1],
Nick Hawes[3], Martijn Wisse[1], and Riccardo Ferrari[2]

[1] Cognitive Robotics, Delft University of Technology, Delft, Netherlands
{c.pezzato,c.h.corbato,m.wisse}@tudelft.nl
[2] Delft Center for Systems and Control, Delft University of Technology,
Delft, Netherlands
r.ferrari@tudelft.nl
[3] Oxford Robotics Institute, University of Oxford, Oxford, UK
{mohamed,nickh}@robots.ox.ac.uk

Abstract. We present a fault tolerant control scheme for robot manipulators based on active inference. The proposed solution makes use of the sensory prediction errors in the free-energy to simplify the residuals and thresholds generation for fault detection and isolation and does not require additional controllers for fault recovery. Results validating the benefits in a simulated 2DOF manipulator are presented and the limitations of the current approach are highlighted.

Keywords: Fault-tolerant control · Fault recovery · Active inference · Free-energy · Robot manipulator

1 Introduction

Developing fault tolerant (FT) control schemes is of vital importance to bring robots outside controlled laboratories. The area of fault tolerant control has become increasingly more important in recent years, and several methods have been developed in different fields. An extensive bibliographical review and classification of FT methods can be found in [25]. Model-based FT techniques are amongst the most promising approaches [8]. For fault detection, they rely on mathematical models to generate *residual* signals to be compared to a *threshold*. Fault recovery is then often performed by switching among different available fault-specific controllers [17]. The two main challenges to design FT schemes are the definition of residuals and thresholds, and the design of a fault specific recovery strategy. We present a novel FT scheme for sensory faults [19,23] based on an active inference controller (AIC) [20], which is inspired by the active inference framework. Active inference is prominent in the neuroscientific literature as a general theory of the brain [9–11] and several recent approaches in robotics have taken inspiration from it [1–3,15,16,18,20–22,24]. In this work we investigate the utility of the active inference framework for fault-tolerant control with sensory faults. In the presented scheme, we exploit the properties of the framework to simplify the definition of both residuals and thresholds, and we provide a simple and general mechanism for sensory fault recovery. Our approach is validated on a simulated 2DOF manipulator.

2 Problem Statement

We consider a robot controlled in joint space with torque commands, using an active inference controller (AIC) [1,20]. In the following, we highlight the necessary elements and assumptions to derive an expression for the free-energy of the system, and the equations for state estimation and control. This study considers a 2-DOF robot arm (Fig. 1), equipped with a vision system to retrieve the end effector Cartesian position $\boldsymbol{y}_v = [y_{v_x},\ y_{v_z}]^\top$, and with position and velocity sensors $\boldsymbol{y}_q,\ \boldsymbol{y}_{\dot{q}} \in \mathbb{R}^2$ for the two joints. Thus, we define $\boldsymbol{y} = [\boldsymbol{y}_q,\ \boldsymbol{y}_{\dot{q}},\ \boldsymbol{y}_v]$. The proprioceptive sensors and the camera are affected by zero mean Gaussian noise $\boldsymbol{\eta} = [\boldsymbol{\eta}_q,\ \boldsymbol{\eta}_{\dot{q}},\ \boldsymbol{\eta}_v]$. Additionally, the camera is affected by barrel distortion. The states $\boldsymbol{x}$ to be controlled are set as the joint positions $\boldsymbol{q}$ of the robot arm. We define the generative model of the state dynamics $\boldsymbol{f}(\cdot)$ such that the robot will be steered to a desired joint configuration $\boldsymbol{\mu}_d$ following the dynamics of a first order linear system with time constant τ.

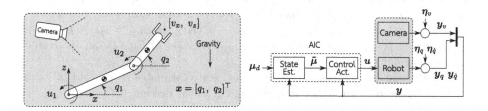

Fig. 1. 2-DOF robot arm and general AIC control scheme.

$$\boldsymbol{f}(\boldsymbol{\mu}) = (\boldsymbol{\mu}_d - \boldsymbol{\mu})\tau^{-1} \tag{1}$$

The relation between $\boldsymbol{\mu}$ and $\boldsymbol{y}$ is expressed through the generative model of the sensory input $\boldsymbol{g} = [\boldsymbol{g}_q,\ \boldsymbol{g}_{\dot{q}},\ \boldsymbol{g}_v]$. Since we set $\boldsymbol{x} = [q_1, q_2]^\top$ and we can directly measure joint positions, $\boldsymbol{g}_q$ and $\boldsymbol{g}_{\dot{q}}$ and their partial derivatives are [7,20]:

$$g_q(\boldsymbol{\mu}) = \boldsymbol{\mu}, \quad g_{\dot{q}}(\boldsymbol{\mu}) = \boldsymbol{\mu}', \quad \partial g_q(\boldsymbol{\mu})/\partial\boldsymbol{\mu} = 1, \quad \partial g_{\dot{q}}(\boldsymbol{\mu})/\partial\boldsymbol{\mu}' = 1 \tag{2}$$

Note that $\boldsymbol{\mu}'$ is the first order generalised motion of $\boldsymbol{\mu}$. To define $g_v(\boldsymbol{\mu})$, instead, we use a *Gaussian Process Regression* (GPR) as in [16]. The training data is composed by a set of observations of the camera output $[\bar{\boldsymbol{y}}_{v_x}, \bar{\boldsymbol{y}}_{v_z}]^\top$ in several robot configurations $\bar{\boldsymbol{y}}_q$. We use a squared exponential kernel k of the form:

$$k(\boldsymbol{y}_{q_i}, \boldsymbol{y}_{q_j}) = \sigma_f^2 \exp\left(-\tfrac{1}{2}(\boldsymbol{y}_{q_i} - \boldsymbol{y}_{q_j})^\top \Lambda (\boldsymbol{y}_{q_i} - \boldsymbol{y}_{q_j})\right) + \sigma_n^2 d_{ij}$$

where $\boldsymbol{y}_{q_i}$, $\boldsymbol{y}_{q_j} \in \bar{\boldsymbol{y}}_q$, and d_{ij} is the Kronecker delta function. Λ is a diagonal matrix of hyperparameters to be optimised. It holds then:

$$g_v(\boldsymbol{y}_{q_*}) = \begin{bmatrix} K_* K^{-1} \bar{\boldsymbol{y}}_{v_x} \\ K_* K^{-1} \bar{\boldsymbol{y}}_{v_z} \end{bmatrix} \quad g_v(\boldsymbol{y}_{q_*})' = \begin{bmatrix} -\Lambda^{-1}(\boldsymbol{y}_{q_*} - \bar{\boldsymbol{y}}_q)^\top [k(\boldsymbol{y}_{q_*}, \bar{\boldsymbol{y}}_q)^\top \cdot \boldsymbol{\alpha}_x] \\ -\Lambda^{-1}(\boldsymbol{y}_{q_*} - \bar{\boldsymbol{y}}_q)^\top [k(\boldsymbol{y}_{q_*}, \bar{\boldsymbol{y}}_q)^\top \cdot \boldsymbol{\alpha}_z] \end{bmatrix} \quad (3)$$

where $\cdot$ means element-wise multiplication, $\boldsymbol{\alpha}_x = K^{-1} \bar{\boldsymbol{y}}_{v_x}$ and $\boldsymbol{\alpha}_z = K^{-1} \bar{\boldsymbol{y}}_{v_z}$, with K being the covariance matrix.

Given the generative models $\boldsymbol{f}$ and $\boldsymbol{g}$ as before, we can define an expression for the free-energy $\mathcal{F}$. Under Laplace approximation, and considering normally distributed uncorrelated noise and generalised motions up to second order, the free-energy for the 2-DOF robot arm can be expressed as:

$$\mathcal{F} = \frac{1}{2} \sum_i \left(\boldsymbol{\varepsilon}_i^\top P_i \boldsymbol{\varepsilon}_i + \ln|P_i|\right) + C, \qquad i \in \{y_q, y_{\dot{q}}, y_v, \mu, \mu'\} \quad (4)$$

where and C is a constant and P_i defines a precision (or inverse covariance) matrix. Note that we set $\tau = 1$ as in [20]. The terms $\boldsymbol{\varepsilon}_i = (\boldsymbol{y}_i - \boldsymbol{g}_i(\mu))$ with $i \in \{y_q, y_{\dot{q}}, y_v\}$ are the *Sensory Prediction Errors* (SPE), representing the difference between observed sensory input and expected one. The model prediction errors are instead defined considering the desired state dynamics as $\boldsymbol{\varepsilon}_\mu = (\mu' - \boldsymbol{f}(\mu))$ and $\boldsymbol{\varepsilon}_{\mu'} = (\mu'' - \partial f(\mu)/\partial_\mu \mu')$. In particular, for the 2-DOF example it results that $\varepsilon_q = (\boldsymbol{y}_q - \mu)$, $\varepsilon_{\dot{q}} = (\boldsymbol{y}_{\dot{q}} - \mu')$, $\varepsilon_v = (\boldsymbol{y}_v - \boldsymbol{g}_v(\mu))$, and $\varepsilon_\mu = (\mu' + \mu - \mu_d)$, $\varepsilon_{\mu'} = (\mu'' + \mu')$. For more details on the derivation of Eq. (4), an interested reader can refer to [6,7,20].

Finally, one can compute the generalised state estimates μ, μ', and μ'', and control actions $\boldsymbol{u}$ by minimizing $\mathcal{F}$ through gradient descent [11]:

$$\dot{\mu} = \mu' - \kappa_\mu \frac{\partial \mathcal{F}}{\partial \mu}, \qquad \dot{\mu}' = \mu'' - \kappa_\mu \frac{\partial \mathcal{F}}{\partial \mu'}, \qquad \dot{\mu}'' = -\kappa_\mu \frac{\partial \mathcal{F}}{\partial \mu''} \quad (5)$$

$$\dot{\boldsymbol{u}} = -\kappa_a \frac{\partial \boldsymbol{y}_q}{\partial \boldsymbol{u}} P_{y_q}(\boldsymbol{y}_q - \mu) - \kappa_a \frac{\partial \boldsymbol{y}_{\dot{q}}}{\partial \boldsymbol{u}} P_{y_{\dot{q}}}(\boldsymbol{y}_{\dot{q}} - \mu') - \kappa_a \frac{\partial \boldsymbol{y}_v}{\partial \boldsymbol{u}} P_{y_v}(\boldsymbol{y}_v - \boldsymbol{g}_v(\mu)) \quad (6)$$

Note that P_{y_q}, $P_{y_{\dot{q}}}$ and P_{y_v} are the precision matrices representing the confidence about sensory inputs. The higher the confidence in a sensor, the more reliable its measurements are assumed to be. Following [18,20], we set $\partial y_q/\partial \boldsymbol{u}$ and $\partial y_{\dot{q}}/\partial \boldsymbol{u}$ to the identity, approximating the true relationships with only their sign. Similar considerations can be made for the relation between commanded torques and Cartesian displacements. The sign of $\partial y_v/\partial \boldsymbol{u}$ depends on the combination of the two joint angles. For instance, operating the end effector in the fourth quadrant with $-\pi/2 \le q_1 \le 0$, a positive u_1 will lead to positive increments of both x and z coordinates. More advanced methods to determine these partial derivatives are out of the scope of this work, but definitely possible and encouraged.

3   A Fault Tolerant Scheme Based on Active Inference

In this section we define a fault tolerant scheme as in Fig. 2, using the SPE to build residuals for fault detection. We also show how the sensory redundancy and precision matrices can be used for fault recovery, such that we do not need to generate extra signals for detection as in conventional approaches, and we simplify fault recovery.

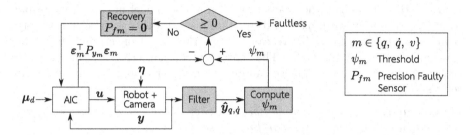

Fig. 2. AIC with additional elements (in gray) for fault tolerant control.

Threshold for Fault Detection and Isolation (FDI). Even though the SPE can be seen as residuals for fault detection purposes, there is a substantial difference. In active inference, the belief μ is biased towards a given goal, so the SPE can also increase for causes which are not related to sensory faults (i.e. the robot is stuck due to a collision). This precludes the use of established fault detection techniques to define a threshold ψ_m for FDI. To solve this issue we consider the quadratic form $\varepsilon_m^\top P_{y_m} \varepsilon_m$, where $\varepsilon_m = y_m - g_m(\mu)$ with $m \in \{q, \dot{q}, v\}$. The core idea is to compute an upper bound on this quadratic term. The first step is to compute an estimate for the prediction errors using the generative model as $\hat{\varepsilon}_m = g_m(\hat{x}) - g_m(\mu)$. The estimate of the joint positions $\hat{x}$ is obtained through a filter using position and velocity measurements. Defining $\dot{x} = z$, we can write:

$$\dot{\hat{x}} = \hat{z} + H_1(y_q - \hat{x}) \qquad \dot{\hat{z}} = H_2(y_{\dot{q}} - \hat{z}) \tag{7}$$

where H_1, H_2 are diagonal positive definite matrices. The estimation error can be made arbitrary small by choosing high gains H_1 and H_2 [12].

We can represent the sensory input y as a function of the ground truth x as:

$$y = g(x) + \gamma + \eta, \tag{8}$$

where $\gamma \in \mathbb{R}^6$ is a vector representing the process uncertainties introduced by the generative models g, and η is the measurement noise. The sensory prediction error for a generic sensor m can then be written as:

$$\varepsilon_m = \hat{\varepsilon}_m + \underbrace{g_m(x) - g_m(\hat{x}) + \gamma_m + \eta_m}_{\delta_m} = \hat{\varepsilon}_m + \delta_m \tag{9}$$

where $\boldsymbol{\delta}_m$ is the total uncertainty including process and measurement noise. The j-th entry $\delta_m(j)$ is a scalar total uncertainty associated with a specific sensor. Since we operate the robot in a finite workspace, with specific physical limits, the states of the system $\boldsymbol{x}$, the sensory input $\boldsymbol{y}$ and the internal belief $\boldsymbol{\mu}$ remain bounded in a compact region $\mathcal{R} = \mathcal{R}_x \times \mathcal{R}_y \times \mathcal{R}_\mu \subset \mathbb{R}^2 \times \mathbb{R}^6 \times \mathbb{R}^2$, before and after the occurrence of a fault. We also suppose that the noise $\boldsymbol{\eta}$ affecting the position and velocity sensors, and the camera, can be bounded. This means $\|\boldsymbol{\eta}(t)\|_2 \leq \bar{\eta}$, where $\bar{\eta}$ is a known value. Since the AIC does not require the full dynamical model of the robot arm, the characterization of the model uncertainties $\boldsymbol{\gamma}$ due to $\boldsymbol{g}(\cdot)$ is straightforward: $\boldsymbol{g}_q(\cdot)$ and $\boldsymbol{g}_{\dot{q}}(\cdot)$ are just an identity, so no uncertainty is introduced, while for $\boldsymbol{g}_v(\cdot)$ we can retrieve the model uncertainty from the covariance matrix of the GPR. The sensory prediction errors $\boldsymbol{\varepsilon}_m$ and $\hat{\boldsymbol{e}}_m$ are then bounded quantities, thus we can define an upper bound for the quadratic term $\boldsymbol{\varepsilon}_m^\top P_{y_m} \boldsymbol{\varepsilon}_m$.

Definition 1. *Given a maximum uncertainty $\bar{\boldsymbol{\delta}}_m$ such that $|\delta_m(j)| \leq \bar{\delta}_m(j) \; \forall j$, we define a threshold for fault detection for sensor m as:*

$$\psi_m = \hat{\boldsymbol{e}}_m^\top P_{y_m} \hat{\boldsymbol{e}}_m + 2|\hat{\boldsymbol{e}}_{y_m}^\top P_{y_m} \bar{\boldsymbol{\delta}}_m| + \bar{\boldsymbol{\delta}}_m^\top P_{y_m} \bar{\boldsymbol{\delta}}_m \tag{10}$$

Lemma 1. *In a faultless case, the quadratic form of the sensory prediction errors for a sensor m will remain below the threshold ψ_m:*

$$\boldsymbol{\varepsilon}_m^\top P_{y_m} \boldsymbol{\varepsilon}_m \leq \psi_m \tag{11}$$

Proof. Once $\bar{\boldsymbol{\delta}}_m$ is given, and since P_{y_m} is diagonal positive definite, Eq. (11) follows from applying the triangular inequality, considering $\boldsymbol{\varepsilon}_m$ as in Eq. (9).

Using Lemma 1, a fault in a generic sensor m is detected and isolated whenever Eq. (11) is violated, that is when a fault will produce an anomaly in the sensory input bigger than $\bar{\boldsymbol{\delta}}_m$. The value for the maximum uncertainty is chosen according to the standard deviation of the noise present in the sensors. Note that, in theory, a bound $\bar{\eta}$ may not be finite or could be very large making fault detection difficult or even impossible. In practice, using multiples of the variance, we reach an acceptable compromise between false alarms and detectability. Thus, each entry of $\bar{\boldsymbol{\delta}}_m$ is set to $5\sigma_m$ where $\boldsymbol{\eta}_m \sim \mathcal{N}(0, \sigma_m^2 I)$. Doing so, the probability of having a false alarm due to the noise is less that 10^{-6}.

Fault Recovery. To recover from a fault we exploit the fact that the controller encodes the precision matrices P_{y_q}, $P_{y_{\dot{q}}}$ and P_{y_v}. Once a fault is detected and isolated, fault recovery can be implemented simply by setting the precision matrix of the faulty sensor to zero, that is $P_{f_m} = \boldsymbol{0}$. This is a simple and generic mechanism to recover for any kind of sensory fault once detected.

4 Simulation Results

We control the robot from the initial position $\boldsymbol{q} = [-\pi/2, \; 0] \; (rad)$ to the desired position $\boldsymbol{\mu}_d = [-0.6, \; 0.5] \; (rad)$. A fault is injected either in the encoders or in the camera during the motion of the robot, at time $t_f = 2$ (s). The maximum uncertainties are set to $\bar{\boldsymbol{\delta}}_q = [5\sigma_q,\ 5\sigma_q]^\top$ and $\bar{\boldsymbol{\delta}}_v = [5\sigma_v,\ 5\sigma_v]^\top$, where $\sigma_q = 0.001$ and $\sigma_v = 0.01$. Figure 3A reports the single normalised SPE $\varepsilon_m^\top \sigma_m^{-1} \varepsilon_m / \psi_m$, that we call $N\varepsilon_m$. Doing so, a fault is detected when the ratio is bigger than one. We assume two kinds of possible faults: 1) A fault in the encoders: the output related to the first joint freezes so $\boldsymbol{y}_q(t) = [q_1(t_f),\ q_2(t)]^\top + \boldsymbol{\eta}_q$ for $t \geq t_f$, and 2) A fault in the camera: a misalignment is injected as bias in $\boldsymbol{y}_v$. Figure 3A shows fault detection and isolation at t_{DI}, when the normalised residual is bigger than 1. The recovered and non-recovered response of the robot in case of encoder fault is depicted in Fig. 3B. A similar response is found for camera faults.

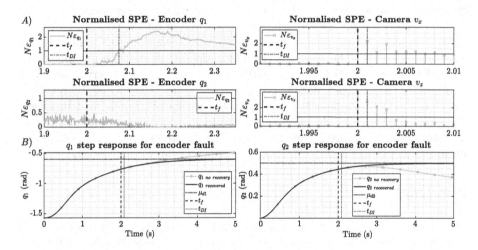

Fig. 3. A) Normalised SPE for FDI in case of encoder fault (left) or camera fault (right). B) Step response with and without recovery action in case of encoder fault.

5   Discussion and Conclusion

Consider now Eq. (1). The time constant τ influences the generative model of the state dynamics $\boldsymbol{f}(\cdot)$, so the desired evolution of the states. As explained in [1], the AIC has two extremes depending on the value of τ^{-1} in the generative model. If $\tau^{-1} \to 0$, the estimation step has zero bias towards the target. The control action in this case will never steer the system towards the target. On the other hand, if $\tau^{-1} \to \infty$ the system is completely biased towards the target. That case is equivalent to a PID controller [1,4,5]. For any value in between, there is a compromise between estimation and control. The estimation and control are thus 'coupled'. This has a few limitations. First, the actions are not explicit in the model, so only sensory faults can be detected, isolated, and recovered. Second, the estimated state is always biased towards the desired state. Finally, biasing the state hinders learning model (hyper-)parameters.

What does this mean for the FT scheme presented so far? The bias could increase the SPE for reasons unrelated to sensory faults, for instance if the current target state changes to another which is further away from the current position. This prevents the direct use of the SPE as residuals in combination with established fault detection techniques, since it would cause several false positives. It would then be beneficial to decouple estimation and control. In addition, a decoupled system could facilitate learning the hyperparameters. This could allow us to optimise the precision matrices for the SPE instead of setting P_{fm} to zero, since the precision matrices would represent the physical noise affecting the sensors. Decoupling can also help relaxing the assumption on the maximum $\bar{\delta}_m$ which now has to be known a priori for the determination of the fault detection threshold. Approaches where the estimation and control are decoupled (similar to [3,13,14,24]) for fault tolerance will be explored in future work.

To conclude, in this paper we present a novel approach for FT control based on active inference. The main novelty is the definition of an on-line threshold for FDI based on the SPE defined in the free-energy. Fault recovery is achieved by reducing the precision of faulty sensor to zero, providing a generic recovery mechanism which significantly simplifies the synthesis of reliable FT controllers. The main limitation of the proposed approach is that only sensory faults can be detected and recovered. Simulation results validated the theoretical findings. Future work will explore FT control with decoupled state-estimation and control.

References

1. Baioumy, M., Duckworth, P., Lacerda, B., Hawes, N.: Active inference for integrated state-estimation, control, and learning. arXiv preprint arXiv:2005.05894 (2020)
2. Baioumy, M., Mattamala, M., Duckworth, P., Lacerda, B., Hawes, N.: Adaptive manipulator control using active inference with precision learning. In: UKRAS (2020)
3. Baioumy, M., Mattamala, M., Hawes, N.: Variational inference for predictive and reactive controllers. In: ICRA 2020 Workshop on New Advances in Brain-Inspired Perception, Interaction and Learning, Paris, France (2020)
4. Baltieri, M., Buckley, C.L.: A probabilistic interpretation of PID controllers using active inference. In: Manoonpong, P., Larsen, J.C., Xiong, X., Hallam, J., Triesch, J. (eds.) SAB 2018. LNCS (LNAI), vol. 10994, pp. 15–26. Springer, Cham (2018). https://doi.org/10.1007/978-3-319-97628-0_2
5. Baltieri, M., Buckley, C.L.: PID control as a process of active inference with linear generative models. Entropy **21**(3), 257 (2019)
6. Bogacz, R.: A tutorial on the free-energy framework for modelling perception and learning. J. Math. Psychol. **76**, 198–211 (2017)
7. Buckley, C.L., Kim, C.S., McGregor, S., Seth, A.K.: The free energy principle for action and perception: a mathematical review. J. Math. Psychol. **81**, 55–79 (2017)
8. Chen, J., Patton, R.J.: Robust Model-Based Fault Diagnosis for Dynamic Systems. Springer, New York (1999). https://doi.org/10.1007/978-1-4615-5149-2
9. Friston, K.J.: The free-energy principle: a unified brain theory? Nat. Rev. Neurosci. **11**(2), 27–138 (2010)

10. Friston, K.J., Daunizeau, J., Kiebel, S.: Action and behavior: a free-energy formulation. Biol. Cybern. **102**(3), 227–260 (2010)
11. Friston, K.J., Mattout, J., Kilner, J.: Action understanding and active inference. Biol. Cybern. **104**(1–2), 137–160 (2011)
12. Khalil, H.K.: High-gain observers in nonlinear feedback control. In: 2008 International Conference on Control, Automation and Systems (2008)
13. van de Laar, T., Özçelikkale, A., Wymeersch, H.: Application of the free energy principle to estimation and control. arXiv preprint arXiv:1910.09823 (2019)
14. van de Laar, T.W., de Vries, B.: Simulating active inference processes by message passing. Front. Robot. AI **6**(20) (2019)
15. Lanillos, P., Cheng, G.: Active inference with function learning for robot body perception. In: International Workshop on Continual Unsupervised Sensorimotor Learning (ICDL-Epirob) (2018)
16. Lanillos, P., Cheng, G.: Adaptive robot body learning and estimation through predictive coding. In: IROS (2018)
17. Narendra, K.S., Balakrishnan, J.: Adaptive control using multiple models. IEEE Trans. Autom. Control (1997)
18. Oliver, G., Lanillos, P., Cheng, G.: Active inference body perception and action for humanoid robots. arXiv preprint arXiv:1906.03022v2 (2019)
19. Paviglianiti, G., Pierri, F., Caccavale, F., Mattei, M.: Robust fault detection and isolation for proprioceptive sensors of robot manipulators. Mechatronics **20**(1), 162–170 (2010)
20. Pezzato, C., Ferrari, R., Corbato, C.H.: A novel adaptive controller for robot manipulators based on active inference. IEEE Robot. Autom. Lett. (2020)
21. Pio-Lopez, L., Nizard, A., Friston, K., Pezzulo, G.: Active inference and robot control: a case study. J. Roy. Soc. Interface **13**(122) (2016)
22. Sancaktar, C., Lanillos, P.: End-to-end pixel-based deep active inference for body perception and action. arXiv preprint arXiv:2001.05847 (2019)
23. Van, M., Wu, D., Ge, S., Ren, H.: Fault diagnosis in image-based visual servoing with eye-in-hand configurations using Kalman filter. IEEE Trans. Ind. Electron. **12**(6), 1998–2007 (2016)
24. Vanderbroeck, M., Baioumy, M., van der Lans, D., de Rooij, R., van der Werf, T.: Active inference for robot control: a factor graph approach. Student Undergraduate Res. E-J. **5**, 1–5 (2019)
25. Zhang, Y., Jiang, J.: Bibliographical review on reconfigurable fault-tolerant control systems. Annu. Rev. Control **32**(2), 229–252 (2008)

A Worked Example of
Fokker-Planck-Based Active Inference

Magnus T. Koudahl$^{(\boxtimes)}$ and Bert de Vries

Electrical Engineering Department, Eindhoven University of Technology,
Groene Loper 19, 5612 AP Eindhoven, The Netherlands
`m.t.koudahl@tue.nl`

Abstract. The Free Energy Principle (FEP) and its corollary active inference describe a complex theoretical framework with a substantial statistical mechanics foundation that is often expressed in terms of the Fokker-Planck equation. Easy-to-follow examples of this formalism are scarce, leaving a high barrier of entry to the field. In this paper we provide a worked example of an active inference agent as a hierarchical Gaussian generative model. We proceed to write its equations of motion explicitly as a Fokker-Planck equation, providing a clear mapping between theoretical accounts of FEP and practical implementation. Code is available at `github.com/biaslab/ai_workshop_2020`.

Keywords: Active inference · Free energy · Fokker-Planck equation

1 Introduction

Theoretical treatments of the free energy principle (FEP) and active inference are often framed in terms of the Fokker-Planck equation [5,6,10,12] and related flows. In this paper we aim to bridge a gap between theory and simulation by providing a worked example of an active inference agent written directly in terms of its Fokker-Planck equation. We provide a brief introduction to the Fokker-Planck description of dynamical systems and implement an agent based on a generative model structure common across the active inference literature. We then successfully apply the agent to a context switching task where it learns to track a harmonic oscillator.

2 The Fokker-Planck Equation for Dynamical Systems

The Fokker-Planck description of dynamical systems [5,6] starts by assuming that the system dynamics can be described by stochastic Langevin equations [5,8] of the form

This work is part of the research programme Efficient Deep Learning with project number P16-25 project 5, which is (partly) financed by the Netherlands Organisation for Scientific Research (NWO).

T. Verbelen et al. (Eds.): IWAI 2020, CCIS 1326, pp. 28–34, 2020.
https://doi.org/10.1007/978-3-030-64919-7_4

$$\dot{\mathbf{x}} = \frac{d\mathbf{x}}{dt} = \mathbf{f}(\mathbf{x}) + \sqrt{2\Gamma(\mathbf{x})}\mathbf{W}(t), \tag{1}$$

where $\mathbf{x}$ denotes the N-dimensional state of the system, $\Gamma(\mathbf{x})$ an $N \times M$ positive semi-definite diffusion matrix and $\mathbf{W}(t)$ is a standard M-dimensional Wiener process. Equation 1 describes the evolution of a system under deterministic state-dependent dynamics $\mathbf{f}(\mathbf{x})$ and a stochastic fluctuation (diffusion) term $\mathbf{W}(t)$. Equivalently, we can consider the time derivative of the distribution generated by Eq. 1 in terms of the Fokker-Planck equation

$$\frac{\partial p(\mathbf{x},t)}{\partial t} = -\sum_{i=1}^{I} \frac{\partial}{\partial x_i} f_i(\mathbf{x})p(\mathbf{x},t) + \sum_{i=1,j=1}^{I,J} \frac{\partial^2}{\partial x_i \partial x_j}\Gamma_{i,j}(\mathbf{x})p(\mathbf{x},t) \tag{2}$$

where both i and j index over dimensions in $\mathbf{x}$. The Fokker-Planck equation describes the time derivative of the distribution $p(\mathbf{x},t)$ generated by Eq. 1 by a deterministic drift component or drag force (the first term) and a random diffusion process (the second term). The core move here is the switch from stochastic realisations of the SDE in Eq. 1 to the deterministic dynamics of the *distribution* over realisations of the same process in Eq. 2. A steady-state solution to the dynamics of Eq. 2 constitutes a vector field and can be written in potential form [1,5,6,8,10] as

$$\mathbf{f}(\mathbf{x}) = (Q(\mathbf{x}) - \Gamma(\mathbf{x}))\nabla J(\mathbf{x}), \tag{3}$$

where $Q(\mathbf{x})$ denotes an anti-symmetric $(Q = -Q^T)$ curl matrix and $\Gamma(\mathbf{x})$ a positive semi-definite diffusion matrix. We use ∇ to denote the gradient and $J(\mathbf{x})$ is a potential function. For a proof of this relation, see [1,8]. Writing $\mathbf{f}(\mathbf{x})$ in this form, the anti-symmetric structure of $Q(\mathbf{x})$ describes a solenoidal flow that is *orthogonal* to gradients of $J(\mathbf{x})$. The positive semi-definiteness of $\Gamma(\mathbf{x})$ on the other hand leads to dissipative flow *along* gradients of $J(\mathbf{x})$.

3   Laplace-Encoded Free Energy and Generative Models

To apply the Fokker-Planck equation to active inference, we follow [5,6] and let $J(\mathbf{x})$ denote a variational free energy functional. We now need to specify a generative model. A common choice in active inference literature [3,6] is a hierarchical generative model of the form

$$
\begin{aligned}
\mu_1 &= h_1(\mu_0) + \omega_1 & \phi_0 &= g_1(\mu_0) + w_0 \\
\mu_2 &= h_2(\mu_1) + \omega_2 & \phi_1 &= g_2(\mu_1) + w_1 \\
&\ \ \vdots & &\ \ \vdots
\end{aligned}
\tag{4}
$$

We let μ_n denote internal states of the agent, ϕ_n sensory states and let $h_n(\cdot)$ and $g_n(\cdot)$ denote arbitrary link functions. Further assuming all noise terms w_n, ω_n

are iid Gaussian, we can rewrite each layer of the hierarchical generative model $p(\mu_{0:n}, \phi_{0:n})$ as

$$p(\mu_{n+1}|\mu_n) = \mathcal{N}(\mu_{n+1}|\mu_n, \sigma^2_{\mu_{n+1}}) = \frac{1}{\sqrt{2\pi\sigma^2_{\mu_{n+1}}}} \exp\left(-\frac{(\mu_{n+1} - h_{n+1}(\mu_n))^2}{2\sigma^2_{\mu_{n+1}}}\right)$$

$$p(\phi_n|\mu_n) = \mathcal{N}(\phi_n|\mu_n, \sigma^2_{\phi_n}) = \frac{1}{\sqrt{2\pi\sigma^2_{\phi_n}}} \exp\left(-\frac{(\phi_n - g_n(\mu_n))^2}{2\sigma^2_{\phi_n}}\right) \qquad (5)$$

where $\sigma^2_{\bullet_n}$ denotes prior variance at the n-th level of $\bullet \in \{\mu, \phi\}$. Once the generative model has been specified, we need to constrain the recognition factors in order to compute the required gradients. Following [3,6] we assume a fully factorised Gaussian recognition density, also known as the mean-field variational Laplace approximation. Under these assumptions the free energy reduces to a sum of precision-weighted prediction errors between internal states at each level μ_n and the level above μ_{n+1}, and internal μ_n and sensory states ϕ_n. This chain can theoretically continue forever. To terminate the chain, we can assume excessive variance (i.e., negligible precision) at the highest level under consideration which renders higher order contributions to the free energy negligible. For a thorough derivation we refer to [2,3]. Ignoring constant terms and summing over levels, the free energy thus takes the form

$$J(\mu_{0:n}, \phi_{0:n}, a) = \sum_n \left(\frac{1}{2\sigma^2_{\mu_{n+1}}}(\mu_{n+1} - h_{n+1}(\mu_n))^2 + \frac{1}{2\sigma^2_{\phi_n}}(\phi_n(a) - g_n(\mu_n))^2\right).$$

$$(6)$$

Note that we additionally assume that ϕ_n depends on active states a. This is the inverse model assumption that augments the generative model [3,4]. The purpose of the inverse model is to update active states by allowing the derivative $\frac{\partial J}{\partial a}$ through

$$\dot{a} = \frac{\partial a}{\partial t} = -\frac{\partial J(\mu_{0:n}, \phi_{0:n})}{\partial a} = -\frac{\partial J(\mu_{0:n}, \phi_{0:n})}{\partial \phi_{0:n}}\frac{\partial \phi_{0:n}}{\partial a} \qquad (7)$$

where we explicitly mediate the effects of action on free energy through the agents sensory states ϕ_n [3,4,6]. This move is usually justified by an appeal to reflex arcs in a neuroscience context [3,4,9] and has successfully been applied in simulation [3,6] as well as robotics [9]. Note that Eq. 7 is effectively a gradient flow on free energy, following the functional form of Eq. 3.

4   A Worked Example

We proceed by defining an environmental process as a harmonic oscillator with an added friction term. The environmental system dynamics are described by a

Hamiltonian (a potential function) that decomposes into potential and kinetic energy terms, plus added friction administered by the agent through action

$$H(x, \dot{x}, a) = \underbrace{\frac{1}{2m}x^2}_{\text{potential}} + \underbrace{\frac{1}{2}k\dot{x}^2}_{\text{kinetic}} - \underbrace{\dot{x}u\tanh(a)}_{\text{friction}}. \tag{8}$$

Here x denotes the position of the system, $\dot{x}$ the velocity, m the mass, k is a constant, u is a force term that bounds the amount of friction the agent can administer and a still represents action. The system obeys standard Hamiltonian dynamics

$$\frac{dx}{dt} = -\frac{\partial H}{\partial \dot{x}}, \frac{d\dot{x}}{dt} = \frac{\partial H}{\partial x}. \tag{9}$$

Hamiltonian dynamics are commonly applied to the description of conservative systems [5]. However in the present example the additional friction term in Eq. 8 means the system no longer conserves energy. In other words, introducing action dependent friction allows the agent to systematically add or subtract energy from the system. If no action is taken ($a = 0$) the third term vanishes and the environmental process describes a standard conservative simple harmonic oscillator.

We let $J(\mu_{0:1}, \phi_0, a)$ denote the free energy of a two-layer model that receives observations only at the first level. The agent thus only observes position and not velocity. Formally this means setting $\phi_0 = x$ and omitting higher orders of ϕ. We then define a new potential vector J' as the concatenation of the Hamiltonian of the environmental process $H(x, \dot{x}, a)$ and the free energy functional $J(\mu_{0:1}, \phi_0, a)$ of the agent

$$J' = \begin{bmatrix} H(x, \dot{x}, a) \\ J(\mu_{0:1}, \phi_0, a) \end{bmatrix} \Rightarrow \nabla J' = \begin{bmatrix} \frac{x}{m} \\ k\dot{x} - u\tanh(a) \\ \frac{1}{\sigma_{\mu_1}^2}(\mu_1 - h_1(\mu_0)) + \frac{1}{\sigma_{\phi_0}^2}(x - g_0(\mu_0)) \\ \frac{1}{\sigma_{\mu_1}^2}(h_1(\mu_0) - \mu_1) \\ -\frac{1}{\sigma_{\phi_0}^2}(x - g_0(\mu_0))u\,\text{sech}^2(a) \end{bmatrix}. \tag{10}$$

We further assume an accurate inverse model for the effect of action a on observations ϕ_0. Concretely this means the agent is able to accurately calculate the gradient flow described in Eq. 7 which is given by

$$\dot{a} = -\frac{\partial J(\mu_{0:1}, \phi_0)}{\partial \phi_0}\frac{\partial \phi_0}{\partial a} = \frac{1}{\sigma_{\phi_0}^2}(x - g_0(\mu_0))u\,\text{sech}^2(a) \tag{11}$$

where $\text{sech}(\cdot)$ is the hyperbolic secant. This derivative can be found in 5th row of $\nabla J'$. Note that the sign is opposite before multiplication by $-\Gamma$. Since the agent does not observe velocity $\dot{x}$, the corresponding sensory prediction error involving ϕ_1 is absent at the second level (4th row of $\nabla J'$).

Concatenating the vector J' allows for simultaneous integration of the agent and the environment by using block matrices for Q and Γ. Assuming

Hamiltonian dynamics for the agent as well, we can write the block system matrices as

$$
Q = \begin{bmatrix} 0 & -1 & 0 & 0 & 0 \\ 1 & 0 & 0 & 0 & 0 \\ 0 & 0 & 0 & -1 & 0 \\ 0 & 0 & 1 & 0 & 0 \\ 0 & 0 & 0 & 0 & 0 \end{bmatrix}, \; \Gamma = \begin{bmatrix} 0 & 0 & 0 & 0 & 0 \\ 0 & 0 & 0 & 0 & 0 \\ 0 & 0 & \gamma_1 & 0 & 0 \\ 0 & 0 & 0 & \gamma_2 & 0 \\ 0 & 0 & 0 & 0 & \gamma_3 \end{bmatrix}, \; \mathbf{x} = \begin{bmatrix} x \\ \dot{x} \\ \mu_0 \\ \mu_1 \\ a \end{bmatrix}. \tag{12}
$$

Here $\mathbf{x}$ denotes the state vector of the combined system similarly to Eq. 3 and Q encodes two blocks of Hamiltonian dynamics. Internal states of the agent additionally perform gradient descent on $J(\mu_{0:1}, \phi_0, a)$ with learning rates γ_1 and γ_2. Action is updated by gradient descent on $J(\mu_{0:1}, \phi_0, a)$ with learning rate γ_3. Note that by virtue of the Fokker-Planck formalism, the learning rates acquire an interpretation in terms of the amplitude of random fluctuations. In other words, maintaining nonequilibrium steady state in a noisy environment mandates high learning rates. Substituting Eq. 10 and 12 into Eq. 3 now finishes the dynamics that underwrite active inference for a partition of states (evolving under the dynamics of Eq. 3) into external states, internal states, sensory states and action.

5   Results

We simulated the system for 50 timesteps with $\gamma_1 = \gamma_2 = 0.1$, $\gamma_3 = 1$, $\sigma_{\bullet n}^2 = 0.1$, $h_n(\mu_n) = g_n(\mu_n) = \mu_n$, $m = k = 1$, $u = 0.5$ and initial state $x = 2$, $\dot{x} = 2$, $\mu_0 = 0$, $\mu_1 = 0$, $a = 0$. Note that the initial states of the agent and the environment are different. At $t = 25$, we change the parameters of the environmental process, setting $m = 10$, $k = 0.1$ and resetting the states of the environment to $x = 10$, $\dot{x} = 2$. This results in an abrupt change in the environmental process. The task of the agent is then twofold: (1) it needs to learn environmental dynamics to accurately predict incoming observations and (2) it needs to flexibly adapt to a change in previously learnt dynamics. Results are shown in Fig. 1. After an initial learning period we observe that the agent accurately learns to track the environmental process. The agents active states settle into an oscillatory pattern to smooth out the trajectory and dampen noise. When the environmental process changes, we observe a new learning period as the agent adapts to the context switch. Prediction errors are quickly attenuated and the agent resumes accurately tracking the environmental process.

6   Discussion

In this paper we showed a worked example of an agent in the form of a common model structure and specified its equations of motion directly in terms of a Fokker-Planck equation. Writing the agent as a Fokker-Planck equation renders the coupling between theory such as [5,6] immediate, with the goal of providing

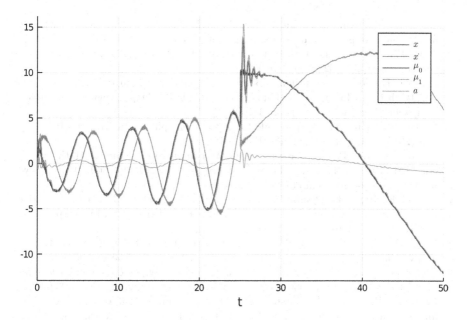

Fig. 1. Trajectory of agent and environmental process. Note the close correspondence between the blue and green lines (x and μ_0), showing that the agent succesfully learns to predict the environmental process. (Color figure online)

an entry point for researchers interested in FEP. A second and more subtle point speaks to the Fokker-Planck equation as a way of writing equations of motion. Writing system dynamics in terms of the Fokker-Planck equation allows for interpreting the equations of motion as a "mechanics" The agent presented here operates under Hamiltonian mechanics but [5] opens up the possibility of investigating quantum- or electro-mechanical agents as well since these can also be written in terms of Q and Γ. Additionally, FEP literature offers a number of alternative free energies that are available as alternatives for the potential function $J(\mathbf{x})$, for example the Expected, Generalised and constrained Bethe free energies [7,11,13]. By combining choices for Q, Γ and $J(\mathbf{x})$ it is immediately clear that Fokker-Planck-based agents represent a sizeable class of agents that are mostly unexplored in practical applications.

References

1. Ao, P.: Potential in stochastic differential equations: novel construction. J. Phys. A Math. General **37**(3), L25–L30 (2004). https://doi.org/10.1088/0305-4470/37/3/L01, http://stacks.iop.org/0305-4470/37/i=3/a=L01?key=crossref.d94e3c0885d8b4979280533ef70299fd
2. Baltieri, M., Buckley, C.: PID control as a process of active inference with linear generative models. Entropy **21**, 257 (2019). https://doi.org/10.3390/e21030257

3. Buckley, C.L., Kim, C.S., McGregor, S., Seth, A.K.: The free energy principle for action and perception: a mathematical review. J. Math. Psychol. **81**, 55–79 (2017). https://doi.org/10.1016/j.jmp.2017.09.004, https://linkinghub.elsevier.com/retrieve/pii/S0022249617300962

4. Friston, K.: What is optimal about motor control? Neuron **72**(3), 488–498 (2011). https://doi.org/10.1016/j.neuron.2011.10.018, http://www.sciencedirect.com/science/article/pii/S0896627311009305

5. Friston, K.: A free energy principle for a particular physics. arXiv:1906.10184 [q-bio] (2019). http://arxiv.org/abs/1906.10184

6. Friston, K., Ao, P.: Free energy, value, and attractors. Comput. Math. Methods Med. **2012**, 1–27 (2012). https://doi.org/10.1155/2012/937860, http://www.hindawi.com/journals/cmmm/2012/937860/

7. Friston, K., Rigoli, F., Ognibene, D., Mathys, C., Fitzgerald, T., Pezzulo, G.: Active inference and epistemic value. Cogn. Neurosci. **6**(4), 187–214 (2015). https://doi.org/10.1080/17588928.2015.1020053

8. Ma, Y.A., Chen, T., Fox, E.B.: A complete recipe for stochastic gradient MCMC. arXiv:1506.04696 [math, stat] (2015). http://arxiv.org/abs/1506.04696

9. Oliver, G., Lanillos, P., Cheng, G.: Active inference body perception and action for humanoid robots. arXiv:1906.03022 [cs] (2019). http://arxiv.org/abs/1906.03022

10. Parr, T., Da Costa, L., Friston, K.: Markov blankets, information geometry and stochastic thermodynamics. Philos. Trans. Roy. Soc. A Math. Phys. Eng. Sci. **378**(2164), 20190159 (2020). https://doi.org/10.1098/rsta.2019.0159, https://royalsocietypublishing.org/doi/10.1098/rsta.2019.0159

11. Parr, T., Friston, K.J.: Generalised free energy and active inference. Biol Cybern. **113**(5–6), 495–513 (2019). https://doi.org/10.1007/s00422-019-00805-w

12. Parr, T., Sajid, N., Friston, K.J.: Modules or mean-fields? Entropy **22**(5), 552 (2020). https://doi.org/10.3390/e22050552, https://www.mdpi.com/1099-4300/22/5/552, number: 5 Publisher: Multidisciplinary Digital Publishing Institute

13. Zhang, D., Wang, W., Fettweis, G., Gao, X.: Unifying message passing algorithms under the framework of constrained bethe free energy minimization. arXiv:1703.10932 [cs, math] (2017). http://arxiv.org/abs/1703.10932

Dynamics of a Bayesian Hyperparameter in a Markov Chain

Martin Biehl[✉] and Ryota Kanai

Araya Inc., Tokyo, Japan
martin@araya.org

Abstract. The free energy principle which underlies active inference attempts to explain the emergence of Bayesian inference in stochastic processes under the assumption of (non-equilibrium) steady state distributions. We contribute a study of the dynamics of an exact Bayesian inference hyperparameter embedded in a Markov chain that infers the dynamics of an observed process. This system does not have a steady-state but still contains exact Bayesian inference. Our study may contribute to future generalizations of the free energy principle to non-steady state systems.

Our treatment uses well-known constructions in Bayesian inference. The main contribution is that we take a different perspective than that of standard treatments. We are interested in how the dynamics of Bayesian inference look from the outside.

Keywords: Free energy principle · Active inference · Markov blankets · Bayesian inference

1 Introduction

One of the most fundamental components of the free energy principle is the approximate Bayesian inference lemma [1]. It claims to provide a sufficient condition for (possibly approximate) Bayesian inference to occur within an ergodic multivariate Markov process. The condition is that there is a partitioning of the variables into internal, active, sensory, and external variables such that the steady-state distribution factorizes in a particular way. If we write μ for internal, a for active, s for sensory, η for external variables and p^* for the steady state density then the required factorization is the conditional independence relation

$$p^*(\mu, \eta|s, a) = p^*(\mu|s, a)p^*(\eta|s, a). \tag{1}$$

This means that (S, A) form a Markov blanket for μ and also for η. However, Bayesian inference can also happen inside processes that don't have steady-state densities. We will illustrate this with two examples below. This explicitly shows that ergodicity and the corresponding Markov blanket condition are only sufficient for Bayesian inference and not necessary.

T. Verbelen et al. (Eds.): IWAI 2020, CCIS 1326, pp. 35–41, 2020.
https://doi.org/10.1007/978-3-030-64919-7_5

Often, the dynamics of the hyperparameters[1] of Bayesian inference are relegated to the background and the focus is on how to compute posteriors for a given hyperparameter or prior. The embedding of both the observed process as well as the hyperparameter into a Markov chain converts standard results into a setting very similar to that of the free energy principle in [1]. The differences are that we have a discrete countably infininte state space instead of a continuous one, discrete instead of continuous time, and in the current version no actions. We will include actions into our setting in future work. Methods for transitioning to continuous systems are well studied so that we are optimistic that insights from the discrete setting can eventually be carried over to the continuous domain. In general we think that the method of embedding Bayesian inference and possibly also approximate Bayesian inference processes into Markov chains can provide rigorously defined examples of interesting systems whose properties can then be studied from an external point of view. The present work exhibits how this can be done in principle.

We observe that the dynamics of the Bayesian hyperparameter can be specified directly in dependence on the last hyperparameter and the observation. This highlights the fact that the probability distributions representing the belief that is being updated are in some sense unnecessary for the dynamics of the process. They have no effect that isn't captured by the hyperparameter itself. This is similar to the situation of the approximate inference lemma where the most likely internal state only "appears" to engage in approximate Bayesian inference with respect to the external state. If we forgot how we derived the hyperparameter dynamics then all we could say is that they appear to engage in Bayesian inference since there is a belief updating process compatible with their dynamics.

2   IID Parameter Inference

Assume as given an identically and independently distributed (IID) random process $(X_t)_{t \in \mathbb{N}}$ with sample space $\mathcal{X}$ specified by a categorical distribution with parameter $\phi = (\phi_x)_{x \in \mathcal{X}}$ which is a vector of the probabilities of the different outcomes i.e. $\phi_x \in [0, 1]$ and

$$\sum_{x \in \mathcal{X}} \phi_x = 1. \tag{2}$$

For each $t \in \mathbb{N}$ we then have

$$p(x_t | \phi) = \prod_x \phi_x^{\delta_{x x_t}} \tag{3}$$

where $\delta_{x x_t}$ is the Kronecker delta.

---

[1] For example, the pseudo-counts that are accumulated as the parameters of a Dirichlet posterior over the categorical states of a generative process.

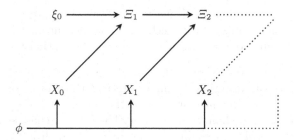

Fig. 1. Bayesian network of the hyperparameter updating process for an IID process with parameter ϕ and initial hyperparameter ξ_0.

We then assume another process $(\Xi_t)_{t\in\mathbb{N}}$ whose dynamics are those of a Bayesian hyperparameter (specifically a parameter of a Dirichlet distribution over parameters of categorical distributions) that updates to parameterize the posterior after each sample from $(X_t)_{t\in\mathbb{N}}$. More precisely, we imagine that for all $t \in \mathbb{N}$ the outcome ξ_t parameterizes a Dirichlet distribution $q(\hat{\phi}|\xi_t)$ over possible (values of/categorical distribution parameters) ϕ.[2] After observing a new sample x_t the posterior $q(\hat{\phi}|x_t, \xi_t)$ is then well defined. To update ξ_t to ξ_{t+1} we require that ξ_{t+1} is the parameter of the posterior. For this we must assume that there exists $\xi \in \Xi$ such that $q(\hat{\phi}|x_t, \xi_t) = q(\hat{\phi}|\xi)$. More generally, we can require that

$$p(\xi_{t+1}|\xi_t, x_t) := \delta_{f(\xi_t, x_t)}(\xi_{t+1}) \tag{4}$$

with

$$f(\xi_t, x_t) := \arg\min_{\xi} \mathrm{KL}[q(\hat{\Phi}|\xi)||q(\hat{\Phi}|x_t, \xi_t)]. \tag{5}$$

In the case we chose where ξ_t is the parameter of a Dirichlet distribution over categorical parameters the solution to this optimisation is[3]

$$\xi_{t+1} = \xi_t + \delta_{x_t} \tag{6}$$

since ξ_t are vectors with $|\mathcal{X}|$ components we can also write this (maybe more clearly) componentwise, i.e. for each component $x \in \mathcal{X}$:

$$(\xi_{t+1})_x := (\xi_t)_x + (\delta_{x_t})_x \tag{7}$$

---

[2] We add a hat to variables that the beliefs encoded by ξ_t range over. This is to highlight that the hatted variables can take different values from the actual ones e.g. when we have a fixed ϕ that defines the IID process then in general the encoded belief $q(\hat{\phi}|\xi_t)$ still ranges over $\hat{\phi} \neq \phi$. A more technical reason is that the hatted variables are in some sense virtual. This should become clearer in the following. A rigorous definition of what "virtual" means is beyond the scope of this paper.

[3] This is the solution because it leads to the KL divergence being zero which means $q(\hat{\Phi}|f(\xi_t, x_t)) = q(\hat{\Phi}|x_t, \xi_t)$. See e.g. [2] for properties of Dirichlet priors for categorical distributions.

Here $(\delta_{x_t})_x := \delta_{x_t x}$ with $\delta_{x_t x}$ the Kronecker delta. In other words, δ_{x_t} is a one-hot encoding of x_t. This defines all the mechanisms/kernels in the Bayesian network Fig. 1 which illustrates our setting. We make two observations:

- Note that while we defined the dynamics of ξ_t via Bayesian inference/updating, the resulting dynamics are just those of a counter of occurrences. There is no reference anymore to the belief $q(\hat{\Phi}|\xi_t)$.
- The resulting Markov chain is not ergodic. The Markov chain state at time t is defined (ξ_t, x_t). A Markov chain can only be ergodic if all states are recurrent. However, since each component $(\xi_t)_x$ of ξ_t is non-decreasing and one of the components increases at every timestep we can never have $\xi_t = \xi_{t+n}$ for any integer $n > 0$.

2.1   Fully Observable Markov Chain

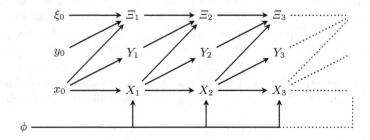

Fig. 2. Bayesian networks of the hyperparameter updating process for a fully observable Markov chain with Markov matrix specified by ϕ, initial Markov chain state x_0, initial stored state y_0, and initial hyperparameter ξ_0. The storage variable Y_t stores the values of X_{t-1} so that Ξ_{t+1} can use the pair $(x_{t-1}, x_t) = (y_t, x_t)$ which indicates the transition that occurred from $t-1$ to t to update.

In order to get an intuition for how to generalise the simple IID case to more interesting cases we look at possibly the next most simple case of inferring the transition probabilities of a time-homogenous Markov chain. Assume as given a time-homogenous finite (discrete-time) Markov chain $(X_t)_{t \in \mathbb{N}}$ with sample space $\mathcal{X}$, initial state $x_0 \in \mathcal{X}$, and Markov matrix (transition probabilities) $p(x_{t+1}|x_t) = \phi_{x_{t+1} x_t}$. Here $(\phi_{x_{t+1} x_t})_{x_{t+1}, x_t \in \mathcal{X}}$ is a matrix of probabilities whose columns sum to one i.e. $\phi_{x_{t+1} x_t} \in [0, 1]$ and for all $x_t \in \mathcal{X}$

$$\sum_{x_{t+1} \in \mathcal{X}} \phi_{x_{t+1} x_t} = 1. \tag{8}$$

For each $t \in \mathbb{N}$ we then have

$$p(x_{t+1}|x_t, \phi) := \prod_{x'x} \phi_{x'x}^{(\delta_{x_{t+1}} \otimes x_t)_{x'x}} \tag{9}$$

where we define for $x, y \in \mathcal{X}$, $\delta_{x \otimes y}$ is a $|\mathcal{X}| \times |\mathcal{X}|$ matrix with

$$(\delta_{x \otimes y})_{ij} := \begin{cases} 1 \text{ if } i = x \text{ and } j = y \\ 0 \text{ else.} \end{cases} \tag{10}$$

We now assume two other processes $(\Xi_t)_{t \in \mathbb{N}}$ and $(Y_t)_{t \in \mathbb{N}}$. Similar to Sect. 2 dynamics of $(\Xi_t)_{t \in \mathbb{N}}$ are those of a Bayesian hyperparameter (a parameter of a Dirichlet distribution over parameters of $|\mathcal{X}|$ categorical distributions) that updates to parameterize the posterior after each transition from x_{t-1} to x_t. Since x_{t-1} is not available to ξ_{t+1} directly it gets stored in y_t. So the update depends on both x_t and y_t. More precisely, we imagine that for all $t \in \mathbb{N}$ the outcome ξ_t parameterizes a Dirichlet distribution $q(\hat{\phi}|\xi_t)$ over possible (values of/categorical distribution parameters) ϕ. At each timestep ξ_{t+1} is updated in response to the pair (y_t, x_t) where x_t is a new sample from the Markov chain and y_t is the stored previous sample x_{t-1} from the Markov chain. At the same time y_{t+1} is updated by setting it equal to x_t. In this way all values/data necessary for the next update are explicitly present at $t + 1$.

The posterior $q(\hat{\phi}|x_t, y_t, \xi_t)$ which corresponds to $q(\hat{\phi}|x_t, x_{t+1}, \xi_t)$ is then well defined. To update ξ_t to ξ_{t+1} we require that ξ_{t+1} is the parameter of the posterior. For this we must assume that there exists $\xi \in \Xi$ such that $q(\hat{\phi}|x_t, y_t, \xi_t) = q(\hat{\phi}|\xi)$. More generally, we can require that

$$p(\xi_{t+1}|\xi_t, x_t, y_t) := \delta_{f(\xi_t, x_t, y_t)}(\xi_{t+1}) \tag{11}$$

with

$$f(\xi_t, x_t, y_t) := \arg\min_{\xi} \mathrm{KL}[q(\hat{\Phi}|\xi) \| q(\hat{\Phi}|x_t, y_t, \xi_t)]. \tag{12}$$

When ξ_t is the parameter of a Dirichlet distribution over categorical parameters this is equivalent to (cmp. [2])

$$f(\xi_t, x_t, y_t) := \xi_t + \delta_{x_t \otimes y_t}. \tag{13}$$

The update of y_t is just the copying of x_t:

$$p(y_{t+1}|x_t) := \delta_{x_t}(y_{t+1}) \tag{14}$$

With this, all the mechanisms/kernels in the Bayesian network Fig. 2 which illustrates our setting are defined.

We make two observations:

- Again the dynamics of ξ_t are just those of a counter of occurrences (of transitions now). There is no reference anymore to a belief.
- The resulting Markov chain is also not ergodic.

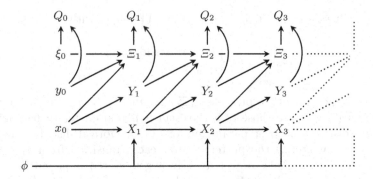

Fig. 3. Bayesian networks of the hyperparameter updating process for a fully observable Markov chain. Here we include the belief distribution as a single random variable Q_t. This random variable takes values in the joint probability distributions over $\hat{\Phi} \times \mathcal{X}^{\mathbb{N}}$ as explained in the text.

Internal Belief Dynamics. We can make the internal belief more explicit by viewing it as a coarse-graining of the Markov chain state (ξ_t, y_t, x_t).

Let $\mathcal{Q}$ be the set of internal belief distributions that each pair (y_t, ξ_t) is mapped to and write Q_t as the random variable that represents the internal belief distribution at time t (see Fig. 3). An instance of such a belief distribution is then denote by q_t.

For a particular (external) timestep t the joint distribution q_t is written

$$q_t(\hat{\phi}, \hat{x}_{-1:\infty}) := \prod_{\tau=0}^{\infty} q(\hat{x}_\tau | \hat{x}_{\tau-1}, \hat{\phi}) q_t(\hat{x}_{t-1}) q_t(\hat{\phi}) \tag{15}$$

where the "intitial" distributions $q_t(\hat{x}_{t-1})$ and $q_t(\hat{\phi})$ will be determined from (y_t, ξ_t) via the two functions we discuss below:

$$q_t(\hat{x}_{t-1}) := b_Y(y_t)(\hat{x}_{-1}) \qquad q_t(\hat{\phi}) := b_\Xi(\xi_t)(\hat{\phi}). \tag{16}$$

First define the functions $b_\Xi : \Xi \to \Delta_{\hat{\phi}}$ and $b_Y : \mathcal{Y} \to \Delta_{\hat{\mathcal{X}}}$ via

$$b_\Xi(\xi)(\hat{\phi}) := \frac{1}{B(\xi)} \prod_{x'x} \hat{\phi}_{x'x}^{(\xi)_{x'x}-1} \qquad b_Y(y)(\hat{x}_{-1}) := \delta_y(\hat{x}_{-1}), \tag{17}$$

where $B(\xi)$ is the beta function.

Using these two distributions as building blocks we can define a third function $b_{Y,\Xi} : \mathcal{Y} \times \Xi \to \Delta_{\Theta \times \mathcal{X}^{\mathbb{N}}}$.

$$b_{Y,\Xi}(y, \xi)(\hat{\phi}, \hat{x}_{-1:\infty}) := \prod_{\tau=0}^{\infty} q(x_\tau | x_{\tau-1}, \hat{\phi}) b_Y(y)(x_{-1}) b_\Xi(\xi)(\hat{\phi}) \tag{18}$$

$$= \frac{1}{B(\xi)} \prod_{x'x} \hat{\phi}_{x'x}^{(\xi+c(x_{-1:\infty}))_{x'x}-1} \delta_y(x_{-1}). \tag{19}$$

where

$$c(x_{0:t}) := \sum_{\tau=1}^{t-1} \delta_{x_\tau} \otimes \delta_{x_\tau - 1}. \tag{20}$$

With this we can define the dependence of Q_t on the external variables (y_t, ξ_t):

$$p(q_t | y_t, \xi_t) := \delta_{b_{Y,\Xi}(y_t, \xi_t)}(q_t). \tag{21}$$

With this the Bayesian network of Fig. 3 is fully defined.

This shows that Q_t for each t is a function of the pair (y_t, ξ_t) and therefore a coarse-graining of the Markov chain state. This highlights the virtual or interpretational nature of the beliefs in this setting. They have no consequence for the next state of the Markov chain.

Acknowledgments. The work by MB and RK on this publication was made possible through the support of a grant from Templeton World Charity Foundation, Inc. The opinions expressed in this publication are those of the authors and do not necessarily reflect the views of Templeton World Charity Foundation, Inc.

References

1. Friston, K.: A free energy principle for a particular physics. arXiv:1906.10184 [q-bio] (2019). http://arxiv.org/abs/1906.10184
2. Minka, T.: Bayesian inference, entropy, and the multinomial distribution. Online Tutorial (2003). https://tminka.github.io/papers/minka-multinomial.pdf

Online System Identification in a Duffing Oscillator by Free Energy Minimisation

Wouter M. Kouw[✉]

Bayesian Intelligent Autonomous Systems Lab,
TU Eindhoven, Eindhoven 5612AP, Netherlands
w.m.kouw@tue.nl

Abstract. Online system identification is the estimation of parameters of a dynamical system, such as mass or friction coefficients, for each measurement of the input and output signals. Here, the nonlinear stochastic differential equation of a Duffing oscillator is cast to a generative model and dynamical parameters are inferred using variational message passing on a factor graph of the model. The approach is validated with an experiment on data from an electronic implementation of a Duffing oscillator. The proposed inference procedure performs as well as offline prediction error minimisation in a state-of-the-art nonlinear model.

Keywords: Online system identification · Duffing oscillator · Free energy minimisation · Variational message passing · Forney factor graphs

1 Introduction

Natural agents are believed to develop an internal model of their motor system by generating actions in muscles and observing limb movements [11]. It has been suggested that forming this internal model is analogous to a form of online system identification [24]. System identification, i.e. estimating dynamical parameters from observed input and output signals, has a rich history in engineering. But there might still be much to gain from considering biologically-plausible procedures. Here, I explore online system identification using a leading theory of how brains process information: free energy minimisation [3,8].

To test free energy minimisation for use in engineering applications, I consider a specific benchmark[1] problem called a Duffing oscillator. Duffing oscillators are relatively well-behaved nonlinear differential equations, making them excellent toy problems for methodological research. Its differential equation is cast to a generative model, with a corresponding factor graph. The factor graph admits a recursive parameter estimation procedure through message passing [12,14]. Specifically, variational message passing minimises free energy [5,13,18]. Here, I infer the parameters of a Duffing oscillator using online variational message passing. Experiments show that it performs as well as a nonlinear ARX model with parameters trained offline using prediction error minimisation [2].

---

[1] http://nonlinearbenchmark.org/.

© Springer Nature Switzerland AG 2020
T. Verbelen et al. (Eds.): IWAI 2020, CCIS 1326, pp. 42–51, 2020.
https://doi.org/10.1007/978-3-030-64919-7_6

2   System

Consider a rigid frame with two prongs facing rightwards (see Fig. 1 left). A steel beam is attached to the top prong. If the frame is driven by a periodic forcing term, the beam will displace horizontally as a driven damped harmonic oscillator. Two magnets are attached to the bottom prong, with the steel beam suspended in between. These act as a nonlinear feedback term on the beam's position, attracting or repelling it as it gets closer [15].

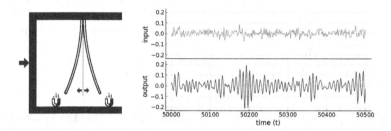

Fig. 1. (Left) Example of a physical implementation of a Duffing oscillator. (Right) Example of input and output signals.

Let $y(t)$ be the observed displacement, $x(t)$ the true displacement, and $u(t)$ the observed driving force. The position of the beam is described as follows [25]:

$$m\frac{d^2x(t)}{dt^2} + c\frac{dx(t)}{dt} + ax(t) + bx^3(t) = u(t) + w(t) \tag{1a}$$

$$y(t) = x(t) + v(t), \tag{1b}$$

where m is mass, c is damping, a the linear and b the nonlinear spring stiffness coefficient. Both the state transition as well as the observation likelihood contain noise terms, which are assumed to be Gaussian distributed: $w(t) \sim \mathcal{N}(0, \tau^{-1})$ (process noise) and $v(t) \sim \mathcal{N}(0, \xi^{-1})$ (measurement noise). The challenge is to estimate m, c, a, b, τ and ξ such that the output of the system can be predicted as accurately as possible.

3   Identification

First, I discretise the state transition of Eq. 1 using a central difference for the second derivative and a forward difference for the first derivative. Re-arranging to form an expression in terms of x_{t+1} yields:

$$x_{t+1} = \frac{2m + c\delta - a\delta^2}{m + c\delta}x_t + \frac{-b\delta^2}{m + c\delta}x_t^3 + \frac{-m}{m + c\delta}x_{t-1} + \frac{\delta^2}{m + c\delta}(u_t + w_t), \tag{2}$$

where δ is the sample time step. Secondly, to ease inference at a later stage, I perform the following variable substitutions:

$$\theta_1 = \frac{2m+c\delta-a\delta^2}{m+c\delta}, \quad \theta_2 = \frac{-b\delta^2}{m+c\delta}, \quad \theta_3 = \frac{-m}{m+c\delta}, \quad \eta = \frac{\delta^2}{m+c\delta}, \quad \gamma = \frac{\tau(m+c\delta)^2}{\delta^4}, \quad (3)$$

where the square in the numerator for γ stems from absorbing the coefficient into the noise term ($\mathbb{V}[\eta w_t] = \eta^2 \mathbb{V}[w_t]$). Note that the mapping between $\phi = (m, c, a, b, \tau)$ and $\psi = (\theta_1, \theta_2, \theta_3, \eta, \gamma)$ can be inverted to recover point estimates:

$$m = \frac{-\theta_3\delta^2}{\eta}, \quad c = \frac{(1+\theta_3)\delta}{\eta}, \quad a = \frac{1-\theta_1-\theta_3}{\eta}, \quad b = \frac{-\theta_2}{\eta}, \quad \tau = \gamma\eta^2. \quad (4)$$

Thirdly, the state transition can be cast to a multivariate first-order form:

$$\underbrace{\begin{bmatrix} x_{t+1} \\ x_t \end{bmatrix}}_{z_t} = \underbrace{\begin{bmatrix} 0 & 0 \\ 1 & 0 \end{bmatrix}}_{S} \underbrace{\begin{bmatrix} x_t \\ x_{t-1} \end{bmatrix}}_{z_{t-1}} + \underbrace{\begin{bmatrix} 1 \\ 0 \end{bmatrix}}_{s} g(\theta, z_{t-1}) + \begin{bmatrix} 1 \\ 0 \end{bmatrix} \eta u_t + \begin{bmatrix} 1 \\ 0 \end{bmatrix} \tilde{w}_t, \quad (5)$$

where $g(\theta, z_{t-1}) = \theta_1 x_t + \theta_2 x_t^3 + \theta_3 x_{t-1}$ and $\tilde{w}_t \sim \mathcal{N}(0, \gamma^{-1})$. The system is now a nonlinear autoregressive process. Lastly, integrating out $\tilde{w}_t$ and v_t produces a Gaussian state transition and a Gaussian likelihood, respectively:

$$z_t \sim \mathcal{N}(f(\theta, z_{t-1}, \eta, u_t), V) \qquad (6a)$$

$$y_t \sim \mathcal{N}(s^\top z_t, \xi^{-1}), \qquad (6b)$$

where $f(\theta, z_{t-1}, \eta, u_t) = S z_{t-1} + s g(\theta, z_{t-1}) + s\eta u_t$ and $V = [\gamma^{-1}\ 0\ ; 0\ \epsilon]$. The number ϵ represents a small noise injection to stabilise inference [6].

To complete the generative model description, priors must be defined. Mass m and process precision τ are known to be strictly positive parameters, while the damping and stiffness coefficients can be both positive and negative. By examining the variable substitutions, it can be seen that θ_1, θ_2, θ_3 and η can be both positive and negative, but γ can only be positive. As such, the following parametric forms can be chosen for the priors:

$$\theta \sim \mathcal{N}(m_\theta^0, V_\theta^0), \quad \eta \sim \mathcal{N}(m_\eta^0, v_\eta^0), \quad \gamma \sim \Gamma(a_\gamma^0, b_\gamma^0), \quad \xi \sim \Gamma(a_\xi^0, b_\xi^0). \quad (7)$$

3.1   Free Energy Minimisation

Given the generative model, a free energy functional with a recognition model q can be formed as follows:

$$-\log p(\mathbf{y}, \mathbf{u}) \leq \iint q(\psi, \mathbf{z}) \frac{q(\psi, \mathbf{z})}{p(\mathbf{y}, \mathbf{u}, \mathbf{z}, \psi)} \, d\mathbf{z} d\psi = \mathcal{F}[q] \qquad (8)$$

where $\mathbf{z} = (z_1, \ldots, z_T)$, $\mathbf{y} = (y_1, \ldots, y_T)$ and $\mathbf{u} = (u_1, \ldots, u_T)$. I assume the states factor over time and that the parameters are largely independent:

$$q(\psi, \mathbf{z}) = q(\theta)q(\eta)q(\gamma)q(\xi) \prod_{t=1}^{T} q(z_t). \qquad (9)$$

All recognition densities are Gaussian distributed, except for $q(\gamma)$ and $q(\xi)$, which are Gamma distributed. In free energy minimisation, the parameters of the recognition distributions depend on each other and are iteratively updated.

3.2 Factor Graphs and Message Passing

In online system identification, parameter estimates should be updated at each time-step. That puts time constraints on the inference procedure. Message passing is an ideal inference procedure due to its efficiency in factorised generative models [12]. Figure 2 is a graphical representation of the generative model, with nodes for factors and edges for variables. Square nodes with Greek letters represent stochastic operations while $\boxed{\cdot}$ and $\boxed{=}$ represent deterministic operations. The node marked "NLARX" represents the state transition described in Eq. 6a.

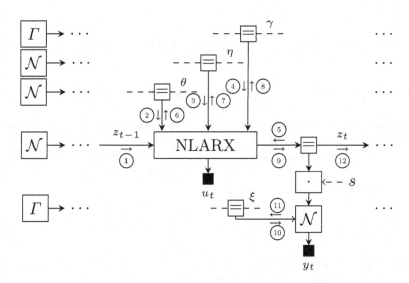

Fig. 2. Forney-style factor graph of the generative model of a Duffing oscillator. Nodes represent conditional distributions and edges represent variables. Nodes send messages to connected edges. When two messages on an edge collide, the marginal belief q for the corresponding variable is updated. Each belief update reduces free energy. By iterating message passing, free energy is minimised.

The terminal nodes on the left represent the initial priors for the states and dynamical parameters. Inference starts when these nodes pass messages. The subgraph - separated by columns of dots - represents the structure of a single time step, recursively applied. Messages ①,②,③,④ and ⑩ represent beliefs q from previous time-steps. Message ⑤, arriving at the state transition node, originates from the likelihood node attached to observation y_t. Messages ⑥,⑦,⑧,⑨ and ⑪ combine priors from previous time steps and likelihoods of observations, and are used to update beliefs q. Message ⑫ is the current state belief and becomes message ① in the next time step.

The graph actually contains more messages, such as those sent by equality nodes. I have hidden them to avoid complicating the figure. Their form has been extensively described in the literature and can be looked up easily [12,14]. Modern message passing toolboxes, such as Infer.NET and ForneyLab.jl, automatically incorporate them. However, the NLARX node is new. Its messages can be computed with[2]:

$$⑥ \; \overrightarrow{\nu}(\theta) = \exp\left(\mathbb{E}_{q(z_t)q(z_{t-1})q(\eta)q(\gamma)}\left[\log \mathcal{N}(f(\theta, z_{t-1}, \eta, u_t), V)\right]\right) \tag{10a}$$

$$⑦ \; \overrightarrow{\nu}(\eta) = \exp\left(\mathbb{E}_{q(z_t)q(z_{t-1})q(\theta)q(\gamma)}\left[\log \mathcal{N}(f(\theta, z_{t-1}, \eta, u_t), V)\right]\right) \tag{10b}$$

$$⑧ \; \overrightarrow{\nu}(\gamma) = \exp\left(\mathbb{E}_{q(z_t)q(z_{t-1})q(\theta)q(\eta)}\left[\log \mathcal{N}(f(\theta, z_{t-1}, \eta, u_t), V)\right]\right) \tag{10c}$$

$$⑨ \; \overrightarrow{\nu}(z_t) = \exp\left(\mathbb{E}_{q(z_{t-1})q(\theta)q(\eta)q(\gamma)}\left[\log \mathcal{N}(f(\theta, z_{t-1}, \eta, u_t), V)\right]\right), \tag{10d}$$

where I use a first-order Taylor expansion to approximate the expected value of the nonlinear autoregressive function $g(\theta, z_{t-1})$.

Loeliger et al. (2007) have written an accessible introduction on message passing in factor graphs [14]. Variational message passing in autoregressive processes has been described in detail as well [5,19].

4   Experiment

The Duffing oscillator has been implemented in an electronic system called Silverbox [25]. It consists of $T = 131702$ samples, gathered with a sampling frequency of 610.35 Hz. Figure 3 shows the time-series, plotted at every 80 time steps. There are two regimes: the first 40000 samples are subject to a linearly increasing amplitude in the input (left of the black line in Fig. 3) and the remaining samples are subject to a constant amplitude but contain only odd harmonics (right of the black line). The second regime is used as a training data set, where both input and output data were given and parameters needed to be inferred. The first regime is used as a validation data set, where the inferred parameters are fixed and the model needs to make predictions for the output signal.

I performed two experiments[3]: a 1-step ahead prediction error and a simulation error setting. I used ForneyLab.jl, with NLARX as a custom node, to run the message passing inference procedure [4]. I call the model above FEM-NLARX,

---

[2] Derivations at https://github.com/biaslab/IWAI2020-onlinesysid.
[3] Experiment notebooks at https://github.com/biaslab/IWAI2020-onlinesysid.

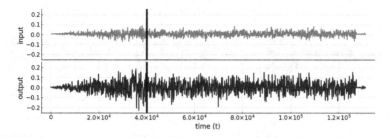

Fig. 3. Silverbox data set, sampled at every 80 time steps for visualisation. The black line splits it into validation data (left) and training data (right).

for Nonlinear Latent Autoregressive model with eXogenous input using Free Energy Minimisation. I implemented two baselines: the first is NLARX without the nonlinearity (i.e. the nonlinear spring coefficient b = 0), dubbed FEM-LARX. The second is a standard NARX model, implemented using MATLAB's System Identification Toolbox. I modelled the static nonlinearity with a sigmoid network of 4 units (in line with the 4 coefficients used by NLARX and LARX). Parameters were inferred offline using Prediction Error Minimisation. Hence, this baseline is called PEM-NARX.

I chose uninformative priors for the coefficients θ and η: Gaussians centred at 1 with precisions of 0.1. The authors of Silverbox indicate that the signal-to-noise ratio at measurement time was high [25]. I therefore chose informative priors for the noise parameters: $a_\xi^0 = 1e8$ and $a_\gamma^0 = 1e3$ (shape parameters) and $b_\xi^0 = 1e3$ and $b_\gamma^0 = 1e1$ (scale parameters).

4.1 1-Step Ahead Prediction Error

At each time-step in the validation data, the models were given the previous output signal y_{t-1}, y_{t-2} and the current input signal u_t and had to infer the current output y_t. It is a relatively easy task, which is reflected in all three models' performance. The top row in Fig. 4 shows the predictions of all three models in purple and their squared error with respect to the true output signal in black. The left column shows the offline NARX baseline (PEM-NARX), the middle column the linear online latent autoregressive baseline (FEM-LARX) and the right column the nonlinear online latent autoregressive model (FEM-NLARX). Note that the errors in the top row seem completely flat. The bottom row in the figure plots the errors on a log-scale. PEM-NARX has a mean squared error of 5.831e−5, FEM-LARX one of 5.945e−5 and FEM-NLARX one of 5.830e−5.

4.2 Simulation Error

In this experiment, the models were not given the previous output signal, but had to use their predictions from the previous time-step. This is a much harder task, because errors will accumulate. The top row in Fig. 5 again shows the predictions

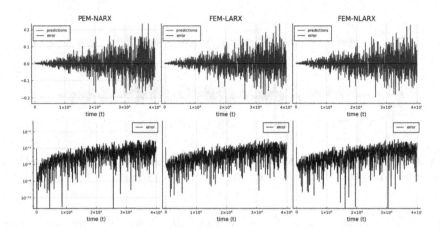

Fig. 4. 1-step ahead prediction errors. (Left) Offline NARX model with sigmoid net (PEM-NARX), (middle) online linear model (FEM-LARX) and (right) online nonlinear model (FEM-NLARX). (Top) Predictions (purple) and squared error (black). (Bottom) Squared prediction errors in log-scale. (Color figure online)

of all three models (purple) and their squared error (black). It can already be seen that the errors increase as the input signal's amplitude rises. The bottom row plots the errors on a log-scale. PEM-NARX has a mean squared error of $1.000e{-}3$, FEM-LARX one of $1.002e{-}3$ and FEM-NLARX one of $0.926e{-}3$.

5   Discussion

The experimental results seem to justify looking to nature for inspiration. Free energy minimisation, in the form of variational message passing, seems a generally applicable and well-performing inference technique. The difficulties mostly lie in deriving variational messages (i.e. Eqs. 10).

Improvements in the proposed procedure could be made with a richer approximation of the nonlinear autoregressive function (e.g. unscented transform) [20]. Alternatively, a hierarchy of latent Gaussian filters or autoregressive processes could be used to obtain time-varying noise parameters or time-varying coefficients [19,22]. Furthermore, instead of discretising such that an auto-regressive model is obtained, one could express the evolution of the states in generalised coordinates. Lastly, black-box models could be explored for further performance improvements.

A natural next step is for an active inference agent to determine the control signal regime (i.e. optimal design). Unfortunately, this is not straightforward: the current formulation relies on variational free energy which does not produce an epistemic term in the objective. The epistemic term is needed to encourage exploration; i.e. try sub-optimal inputs to reduce uncertainty. To arrive at an epistemic term, one would need to work with expected free energy [17]. But it is unclear how expected free energy could be incorporated into factor graphs.

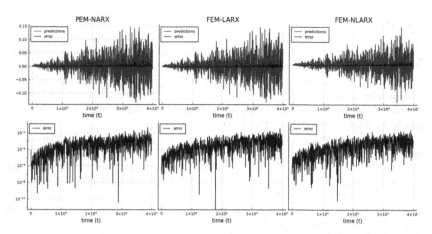

Fig. 5. Simulation errors. (Left) Offline NARX model with sigmoid net (PEM-NARX), (middle) online linear model (FEM-LARX) and (right) online nonlinear model (FEM-NLARX). (Top) Predictions (purple) and squared error (black). (Bottom) Squared prediction errors in log-scale. (Color figure online)

5.1 Related Work

Online system identification procedures typically employ recursive least-squares or maximum likelihood inference, with nonlinearities modelled by basis expansions or neural networks [7,16,23]. Online Bayesian identification procedures come in two flavours: sequential Monte Carlo samplers [1,10] and online variational Bayes [9,26]. This work is novel in the use of variational message passing as an efficient implementation of online variational Bayes and its application to a nonlinear autoregressive model.

6   Conclusion

I have presented a free energy minimisation procedure for online system identification. Experimental results showed comparable performance to a state-of-the-art nonlinear model with parameters estimated offline. This indicates that the procedure performs well enough to be deployed in engineering applications.

Future work should test variational message passing in more challenging nonlinear identification settings, such as a Wiener-Hammerstein benchmark [21]. Furthermore, problems with time-varying dynamical parameters, such as a robotic arm picking up objects with mass, would be interesting for their connection to natural agents.

Acknowledgements. The author thanks Magnus Koudahl, Albert Podusenko and Thijs van de Laar for insightful discussions and the reviewers for their constructive feedback.

References

1. Abdessalem, A.B., Dervilis, N., Wagg, D., Worden, K.: Identification of nonlinear dynamical systems using approximate Bayesian computation based on a sequential Monte Carlo sampler. In: International Conference on Noise and Vibration Engineering (2016)
2. Aguirre, L.A., Letellier, C.: Modeling nonlinear dynamics and chaos: a review. Math. Problems Eng. **2009** (2009)
3. Buckley, C.L., Kim, C.S., McGregor, S., Seth, A.K.: The free energy principle for action and perception: a mathematical review. J. Math. Psychol. **81**, 55–79 (2017)
4. Cox, M., van de Laar, T., de Vries, B.: Forneylab.jl: Fast and flexible automated inference through message passing in Julia. In: International Conference on Probabilistic Programming (2018)
5. Dauwels, J.: On variational message passing on factor graphs. In: IEEE International Symposium on Information Theory, pp. 2546–2550 (2007)
6. Dauwels, J., Eckford, A., Korl, S., Loeliger, H.A.: Expectation maximization as message passing - Part I: Principles and Gaussian messages (2009). arXiv:0910.2832
7. Engel, Y., Mannor, S., Meir, R.: The kernel recursive least-squares algorithm. IEEE Trans. Signal Process. **52**(8), 2275–2285 (2004)
8. Friston, K., Kilner, J., Harrison, L.: A free energy principle for the brain. J. Physiol. **100**(1–3), 70–87 (2006)
9. Fujimoto, K., Satoh, A., Fukunaga, S.: System identification based on variational Bayes method and the invariance under coordinate transformations. In: IEEE Conference on Decision and Control and European Control Conference, pp. 3882–3888 (2011)
10. Green, P.L.: Bayesian system identification of a nonlinear dynamical system using a novel variant of simulated annealing. Mech. Syst. Signal Process. **52**, 133–146 (2015)
11. de Klerk, C.C., Johnson, M.H., Heyes, C.M., Southgate, V.: Baby steps: Investigating the development of perceptual-motor couplings in infancy. Develop. Sci. **18**(2), 270–280 (2015)
12. Korl, S.: A factor graph approach to signal modelling, system identification and filtering. Ph.D. thesis, ETH Zurich (2005)
13. van de Laar, T., Cox, M., Senoz, I., Bocharov, I., de Vries, B.: ForneyLab: a toolbox for biologically plausible free energy minimization in dynamic neural models. In: Conference on Complex Systems (2018)
14. Loeliger, H.A., Dauwels, J., Hu, J., Korl, S., Ping, L., Kschischang, F.R.: The factor graph approach to model-based signal processing. Proc. IEEE **95**(6), 1295–1322 (2007)
15. Moon, F., Holmes, P.J.: A magnetoelastic strange attractor. J. Sound Vibration **65**(2), 275–296 (1979)
16. Paleologu, C., Benesty, J., Ciochina, S.: A robust variable forgetting factor recursive least-squares algorithm for system identification. IEEE Signal Process. Lett. **15**, 597–600 (2008)
17. Parr, T., Friston, K.J.: Generalised free energy and active inference. Biol. Cybern. **113**(5–6), 495–513 (2019)
18. Parr, T., Markovic, D., Kiebel, S.J., Friston, K.J.: Neuronal message passing using mean-field, Bethe, and marginal approximations. Sci. Rep. **9**(1), 1–18 (2019)
19. Podusenko, A., Kouw, W.M., de Vries, B.: Online variational message passing in hierarchical autoregressive models. In: IEEE International Symposium on Information Theory, pp. 1343–1348 (2020)

20. Särkkä, S.: Bayesian Filtering And Smoothing. Cambridge University Press, Cambridge (2013). Vol. 3
21. Schoukens, M., Noël, J.P.: Three benchmarks addressing open challenges in nonlinear system identification. IFAC-PapersOnline **50**(1), 446–451 (2017)
22. Senoz, I., Podusenko, A., Kouw, W.M., de Vries, B.: Bayesian joint state and parameter tracking in autoregressive models. In: Conference on Learning for Dynamics and Control, pp. 1–10 (2020)
23. Tangirala, A.K.: Principles of System Identification: Theory And Practice. CRC Press, Boca Raton (2018)
24. Tin, C., Poon, C.S.: Internal models in sensorimotor integration: perspectives from adaptive control theory. J. Neural Eng. **2**(3), S147 (2005)
25. Wigren, T., Schoukens, J.: Three free data sets for development and benchmarking in nonlinear system identification. In: European Control Conference (ECC), pp. 2933–2938 (2013)
26. Yoshimoto, J., Ishii, S., Sato, M.: System identification based on online variational Bayes method and its application to reinforcement learning. In: Artificial Neural Networks and Neural Information Processing, pp. 123–131. Springer (2003)

Hierarchical Gaussian Filtering
of Sufficient Statistic Time Series
for Active Inference

Christoph Mathys[1,2,3]([✉]) [iD] and Lilian Weber[3,4] [iD]

[1] Interacting Minds Centre, Aarhus University, Aarhus, Denmark
chmathys@cas.au.dk
[2] Scuola Internazionale Superiore di Studi Avanzati (SISSA), Trieste, Italy
[3] Translational Neuromodeling Unit (TNU), Institute for Biomedical Engineering,
University of Zurich and ETH Zurich, Zurich, Switzerland
[4] University of Oxford, Oxford, UK

Abstract. Active inference relies on state-space models to describe the
environments that agents sample with their actions. These actions lead
to state changes intended to minimize future surprise. We show that
surprise minimization relying on Bayesian inference can be achieved by
filtering of the sufficient statistic time series of exponential family input
distributions, and we propose the hierarchical Gaussian filter (HGF) as
an appropriate, efficient, and scalable tool for active inference agents to
achieve this.

Keywords: Active inference · Exponential families · Message passing ·
Precision-weighted prediction errors · Hierarchical Gaussian filter

1 Introduction

Active inference [3] is a framework for modelling and programming the behaviour
of agents negotiating their continued existence in a given environment. Under
active inference, an agent chooses its actions such that they minimize the free
energy of its model of the environment. In order to do this, the agent needs to
perform inference on the state of the environment and its own internal control
states which generate actions.

The agent performing active inference and the researcher modelling such an
agent have a converging interest in a simple, modular, and automated algorithm
that allows them to perform free energy minimization with complex hierarchical models. Accordingly, there have recently been advances in developing an
automated algorithmic framework for free energy minimization in active inference [1,7].

In this paper, we are concerned with the filtering of environmental input
which reaches the agent through its Markov blanket. We show that exponential-
family input distributions can be inferred by tracking the mean of the sufficient statistics of the inputs by passing simple update messages which amount

© Springer Nature Switzerland AG 2020
T. Verbelen et al. (Eds.): IWAI 2020, CCIS 1326, pp. 52–58, 2020.
https://doi.org/10.1007/978-3-030-64919-7_7

to precision-weighted prediction errors. For stationary input distributions, this implements exact Bayesian inference. In the more common case of non-stationary input distributions, we propose to apply hierarchical Gaussian filtering [4,5] to the sufficient statistic time series, resulting in approximate Bayesian inference with a dynamic learning rate.

2   Bayesian Inference Reduced to Mean-Tracking

2.1   Mean Tracking and Exponential Weighting

As a preliminary, we note that the arithmetic mean $\bar{x}_n := \frac{1}{n} \sum_{i=1}^n x_i$ of a time series $\{x_1, x_2, \ldots, x_n\}$ can be updated sequentially from $\bar{x}_n$ to $\bar{x}_{n+1}$ when a new observation x_{n+1} occurs.

$$\bar{x}_{n+1} = \bar{x}_n + \frac{1}{n+1} \left(x_{n+1} - \bar{x}_n \right) \tag{1}$$

If we take the previous mean $\bar{x}_n$ to be a prediction for the new observation x_{n+1}, then the difference $x_{n+1} - \bar{x}_n$ is a *prediction error*. The update to $\bar{x}_n$ then amounts to adding the prediction error weighted by $1/(n+1)$. As n grows, the weight of prediction errors approaches zero, which ensures the equal weighting of all observations in the mean.

As a further preliminary, we note that if we replace the weight $1/(n+1)$ of the prediction error with a constant *learning rate* $\alpha \in [0, 1]$, we no longer get the mean $\bar{x}_n$ of the time series but the exponentially weighted average q_n.

$$q_{n+1} = q_n + \alpha \left(x_{n+1} - q_n \right) \tag{2}$$

With $q_0 := 0$ and $\gamma := 1 - \alpha$, this can be written in closed form,

$$q_n = (1 - \gamma) \sum_{i=0}^{n-1} \gamma^i x_{n-i}, \tag{3}$$

which makes apparent the exponential downweighting of observations x_i as they lie further in the past.

2.2   A Conjugate Prior Which Reduces Bayesian Inference to Mean Tracking for Exponential Families

Exponential families of probability distributions are those which can be written in the form

$$p\left(\boldsymbol{x} | \boldsymbol{\vartheta} \right) = f_{\boldsymbol{x}}(\boldsymbol{\vartheta}) := h(\boldsymbol{x}) \exp \left(\boldsymbol{\eta}(\boldsymbol{\vartheta}) \cdot \boldsymbol{t}(\boldsymbol{x}) - b(\boldsymbol{\vartheta}) \right), \tag{4}$$

where $\boldsymbol{x}$ is a (possibly) vector-valued observation, $\boldsymbol{\vartheta}$ is a parameter vector, $h(\boldsymbol{x})$ is a normalization constant, $\boldsymbol{\eta}(\boldsymbol{\vartheta})$ is the so-called 'natural' parameter vector,

$t(\boldsymbol{x})$ is the sufficient statistic vector, and $b(\boldsymbol{\vartheta})$ is a scalar function. If we choose as our prior

$$p\left(\boldsymbol{\vartheta}|\boldsymbol{\xi},\nu\right) = g_{\boldsymbol{\xi},\nu}(\boldsymbol{\vartheta}) := z\left(\boldsymbol{\xi},\nu\right)\exp\left(\nu\left(\boldsymbol{\eta}\left(\boldsymbol{\vartheta}\right)\cdot\boldsymbol{\xi} - b(\boldsymbol{\vartheta})\right)\right), \tag{5}$$

where $\boldsymbol{\xi}$ is a hyperparameter vector, $\nu > 0$ a scalar hyperparameter, and $z(\boldsymbol{\xi},\nu)$ the normalization constant

$$z(\boldsymbol{\xi},\nu) := \left(\int \exp\left(\nu\left(\boldsymbol{\eta}\left(\boldsymbol{\vartheta}\right)\cdot\boldsymbol{\xi} - b(\boldsymbol{\vartheta})\right)\right)\mathrm{d}\boldsymbol{\vartheta}\right)^{-1}, \tag{6}$$

then the posterior has the same form as the prior (i.e., it is *conjugate*) with updated hyperparameters

$$\nu \leftarrow \nu + 1 \tag{7}$$

$$\boldsymbol{\xi} \leftarrow \boldsymbol{\xi} + \frac{1}{\nu + 1}\left(t(\boldsymbol{x}) - \boldsymbol{\xi}\right). \tag{8}$$

A proof of this is in the Appendix.

In other words, with the prior introduced in Eq. 5, Bayesian inference with exponential family models reduces to tracking the mean of the sufficient statistic $t(\boldsymbol{x}_i)$ of the observations $\{\boldsymbol{x}_1, \boldsymbol{x}_2, \ldots\}$. For a single observation $\boldsymbol{x}$, inference amounts to updating the hyperparameter $\boldsymbol{\xi}$ with the sufficient statistic $t(\boldsymbol{x})$ under the assumption that there have been ν previous observations with sufficient statistic $\boldsymbol{\xi}$.

3   Predictive Distributions

Agents performing active inference minimize the free energy of their model of the environment by minimizing prediction errors regarding their observations (in the long run; in the short run, it is necessary to risk surprises that won't kill us in order to gain the information needed to avoid being dead in the long run). Therefore, the decisive goal and outcome of model-based inference is the *predictive distribution* $\hat{f}$ of inputs $\boldsymbol{x}$. In the present framework, this is

$$\hat{f}_{\boldsymbol{\xi},\nu}(\boldsymbol{x}) := \int f_{\boldsymbol{x}}(\boldsymbol{\vartheta})g_{\boldsymbol{\xi},\nu}(\boldsymbol{\vartheta})\mathrm{d}\boldsymbol{\vartheta}. \tag{9}$$

For the univariate Gaussian with unknown mean and precision, we will call this the *Gaussian-predictive* distribution $\mathcal{NP}$:

$$\hat{f}_{\boldsymbol{\xi},\nu}(x) = \mathcal{NP}\left(x;\boldsymbol{\xi},\nu\right)$$

$$:= \sqrt{\frac{1}{\pi(\nu+1)\left(\xi_{x^2} - \xi_x^2\right)}\frac{\Gamma\left(\frac{\nu+2}{2}\right)}{\Gamma\left(\frac{\nu+1}{2}\right)}} \tag{10}$$

$$\left(1 + \frac{(x - \xi_x)^2}{(\nu+1)\left(\xi_{x^2} - \xi_x^2\right)}\right)^{-\frac{\nu+2}{2}}$$

For $\xi_x = 0$ and $\xi_{x^2} = 1$, this becomes a Student's-t distribution with $\nu+1$ degrees of freedom. Figure 1 shows how the Gaussian-predictive distribution $\mathcal{NP}$ works in practice, i.e., how it adapts as ν, ξ_x, and ξ_{x^2} are updated sequentially according to Eqs. 7 and 8.

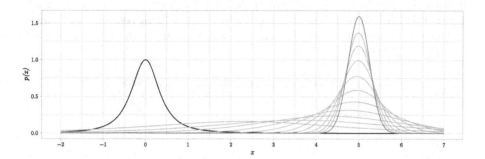

Fig. 1. Sequential updates to the Gaussian-predictive distribution $\mathcal{NP}$ in response to 1024 samples drawn from a Gaussian with mean 5 and standard deviation $1/4$ (red). Initial hyperparameters were $\xi_x = 0$, $\xi_{x^2} = 1/8$, and $\nu = 1$, corresponding to the initial $\mathcal{NP}$ in black. Updated predictive distributions after $2, 4, 8, \ldots, 1024$ samples are shown in grey. (Color figure online)

4 Filtering of Sufficient Statistics for Non-stationary Input Distributions

Active inference agents find themselves in environments where the distributions underlying their observations are non-stationary. In such a setting, older observations have less value for inference about the present than newer ones. Using the hyperparameter update scheme introduced above is then inappropriate because it leads to predictive distributions which rely on outdated information and are overconfident because they overestimate the amount of good information they have. However, since our update scheme relies on tracking the mean of the sufficient statistics of the observations, that is, simply on filtering the sufficient statistic time series, we can apply any known filtering method to this time series and use its output to construct predictive distributions. For example, instead of applying Eq. 1, we could use Eq. 2, which amounts to an exponential downweighting of observations into the past. Using a constant learning rate in this way corresponds to holding ν constant in Eq. 8. As is evident from Eq. 10, this means that the predictive distribution retains its fat tails, meaning that an agent will experience much less surprise at observations far from the predictive mean. However, keeping ν constant raises the question what value to choose for it, and when to change it.

A solution to this is the application of a hierarchical Gaussian filter (HGF) [4,5] to the sufficient statistic time series. The HGF, which contains the Kalman filter as a special case, allows for filtering with an adaptive learning rate which is adjusted according to a continually updated prediction about the volatility of the environment. Updates in the HGF are precision-weighted prediction errors derived from a hierarchical volatility model by variational approximation. For example, in the case of a Gaussian input distribution as in Fig. 1, input x would be filtered by an HGF, allowing for a posterior predictive distribution that dynamically adapts to a volatile input distribution. Figure 2 shows an example of how this procedure yields an adaptive ν, which falls in response to changes in the input distribution and so ensures that the predictive distribution remains fat-tailed at all times.

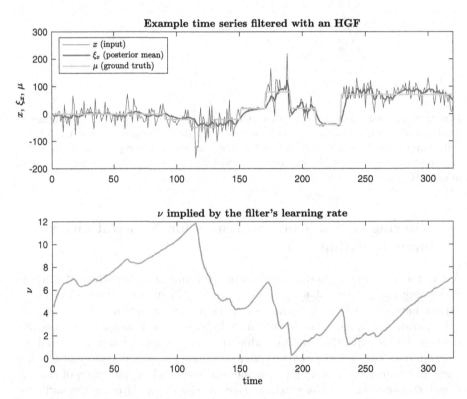

Fig. 2. Example of a time series (input x, top panel, fine blue line) filtered with an HGF (posterior mean ξ_x, top panel, red line), which infers the ground truth μ (top panel, yellow line) well in a volatile environment. Comparison of the HGF updates with Eq. 8 yields implied ν (bottom panel). This never rises above 12, ensuring a fat-tailed predictive distribution. In stable phases, implied ν rises; in volatile phases, it falls. (Color figure online)

5   Discussion

We have shown a way to do exact Bayesian inference with exponential-family models simply by tracking the mean of the sufficient statistic function as observations occur. For this to work, the prior introduced in Eq. 5 is crucial, but its significance has not been recognized before. The approach introduced here is novel. While our prior appears in [2] and seems to have been forgotten since, the resulting updates are there written in a form that obscures their meaning as (precision-)weighted prediction errors and makes it obvious that the relation to mean-tracking was not seen. However, once this is apparent, it supports a *filtering perspective on hyperparameter updates,* which opens up new possibilities such as the HGF filtering proposed in Sect. 4. Additionally, our prior has the benefit of a ready interpretation: ν virtual previous observations with sufficient statistic $\boldsymbol{\xi}$.

For active inference agents, it is critical to predict observations in a way that allows for non-stationary generative processes in the environment. In the framework we propose, this can be achieved by filtering the sufficient statistics of the input distribution using an HGF. This allows predictive distributions to keep a shape (precise but fat-tailed and able to adapt quickly in response to prediction errors) that optimally serves the purpose of minimizing surprise in the long run.

This perspective can be expanded to include networks of HGF nodes where the input distribution and its associated filter are the window into the deeper layers of the network. These deeper layers encode the agent's model of its environment, and it is the free energy of this model that the agent endeavours to minimize by active inference. The present work is therefore a natural complement to recent work on an automated algorithmic framework for free energy minimization in active inference [1,6,7]. The simple message-passing nature of the hyperparameter updates we are proposing fits naturally into message passing schemes in deep networks.

Appendix: Proof of Eqs. 7 and 8

By Bayes' theorem we have

$$
\begin{aligned}
p\left(\boldsymbol{\vartheta}|\boldsymbol{x},\boldsymbol{\xi},\nu\right) &\propto p\left(\boldsymbol{x}|\boldsymbol{\vartheta}\right)p\left(\boldsymbol{\vartheta}|\boldsymbol{\xi},\nu\right)\\
&= f_{\boldsymbol{x}}(\boldsymbol{\vartheta})g_{\boldsymbol{\xi},\nu}(\boldsymbol{\vartheta})\\
&= h(\boldsymbol{x})\exp\left(\boldsymbol{\eta}(\boldsymbol{\vartheta})\cdot\boldsymbol{t}(\boldsymbol{x})-b(\boldsymbol{\vartheta})\right)\\
&\quad z\left(\boldsymbol{\xi},\nu\right)\exp\left(\nu\left(\boldsymbol{\eta}\left(\boldsymbol{\vartheta}\right)\cdot\boldsymbol{\xi}-b(\boldsymbol{\vartheta})\right)\right)\\
&\propto \exp\left(\boldsymbol{\eta}(\boldsymbol{\vartheta})\cdot\left(\boldsymbol{t}(\boldsymbol{x})+\nu\boldsymbol{\xi}\right)-(\nu+1)b(\boldsymbol{\vartheta})\right)
\end{aligned}
$$

We only need to prove that the argument of the exponential function has the required form. Normalization takes care of the rest. Rearranging the argument gives us

$$
\begin{aligned}
\boldsymbol{\eta}(\boldsymbol{\vartheta})&\cdot\left(\boldsymbol{t}(\boldsymbol{x})+\nu\boldsymbol{\xi}\right)-(\nu+1)b(\boldsymbol{\vartheta})\\
&= (\nu+1)\left(\boldsymbol{\eta}(\boldsymbol{\vartheta})\cdot\frac{1}{\nu+1}\left(\boldsymbol{t}(\boldsymbol{x})+\nu\boldsymbol{\xi}\right)-b(\boldsymbol{\vartheta})\right)\\
&= (\nu+1)\left(\boldsymbol{\eta}(\boldsymbol{\vartheta})\cdot\left(\boldsymbol{\xi}+\frac{1}{\nu+1}\left(\boldsymbol{t}(\boldsymbol{x})-\boldsymbol{\xi}\right)\right)-b(\boldsymbol{\vartheta})\right).
\end{aligned}
$$

From this, it follows that

$$p\left(\boldsymbol{\vartheta}|\boldsymbol{x},\boldsymbol{\xi},\nu\right) = g_{\boldsymbol{\xi}',\nu'}\left(\boldsymbol{\vartheta}\right)$$

with

$$\nu' = \nu + 1$$
$$\boldsymbol{\xi}' = \boldsymbol{\xi} + \frac{1}{\nu+1}\left(\boldsymbol{t}(\boldsymbol{x}) - \boldsymbol{\xi}\right)$$

References

1. de Vries, B., Friston, K.J.: A factor graph description of deep temporal active inference. Front. Comput. Neurosci. **11** (2017). https://doi.org/10.3389/fncom.2017.00095
2. Diaconis, P., Ylvisaker, D.: Conjugate priors for exponential families. Ann. Stat. **7**(2), 269–281 (1979)
3. Friston, K.J., Daunizeau, J., Kiebel, S.J.: Reinforcement learning or active inference? PLoS ONE **4**(7), e6421 (2009). https://doi.org/10.1371/journal.pone.0006421
4. Mathys, C., Daunizeau, J., Friston, K.J., Stephan, K.E.: A Bayesian foundation for individual learning under uncertainty. Front. Hum. Neurosci. **5**, 39 (2011). https://doi.org/10.3389/fnhum.2011.00039
5. Mathys, C., et al.: Uncertainty in perception and the Hierarchical Gaussian Filter. Front. Hum. Neurosci. **8**, 825 (2014). https://doi.org/10.3389/fnhum.2014.00825
6. Şenöz, İ., de Vries, B.: Online variational message passing in the hierarchical Gaussian filter. In: 2018 IEEE 28th International Workshop on Machine Learning for Signal Processing (MLSP), pp. 1–6 (Sep 2018). https://doi.org/10.1109/MLSP.2018.8517019
7. van de Laar, T.W., de Vries, B.: Simulating active inference processes by message passing. Front. Robot. AI **6** (2019). https://doi.org/10.3389/frobt.2019.00020

Active Inference and Machine Learning

Deep Active Inference for Partially Observable MDPs

Otto van der Himst[(✉)] and Pablo Lanillos

Department of Artificial Intelligence Donders Insitute for Brain,
Cognition and Behaviour Radboud University,
Montessorilaan 3, 6525 Nijmegen, HR, The Netherlands
o.vanderhimst@student.ru.nl, p.lanillos@donders.ru.nl

Abstract. Deep active inference has been proposed as a scalable approach to perception and action that deals with large policy and state spaces. However, current models are limited to fully observable domains. In this paper, we describe a deep active inference model that can learn successful policies directly from high-dimensional sensory inputs. The deep learning architecture optimizes a variant of the expected free energy and encodes the continuous state representation by means of a variational autoencoder. We show, in the OpenAI benchmark, that our approach has comparable or better performance than deep Q-learning, a state-of-the-art deep reinforcement learning algorithm.

Keywords: Deep active inference · Deep learning · POMDP · Control as inference

1 Introduction

Deep active inference (dAIF) [1–6] has been proposed as an alternative to Deep Reinforcement Learning (RL) [7,8] as a general scalable approach to perception, learning and action. The active inference mathematical framework, originally proposed by Friston in [9], relies on the assumption that an agent will perceive and act in an environment such as to minimize its free energy [10]. Under this perspective, action is a consequence of top-down proprioceptive predictions coming from higher cortical levels, i.e., motor reflexes minimize prediction errors [11].

On the one hand, works on dAIF, such as [2,12,13], have focused on scaling the optimization of the Variational Free-Energy bound (VFE), as described in [9,14], to high-dimensional inputs such as images, modelling the generative process with deep learning architectures. This type of approach preserves the optimization framework (i.e., dynamic expectation maximization [15]) under the Laplace approximation by exploiting the forward and backward passes of

1st International Workshop on Active inference, European Conference on Machine Learning (ECML/PCKDD 2020).

the neural network. Alternatively, pure end-to-end solutions to VFE optimization can be achieved by approximating all the probability density functions with neural networks [1,3].

On the other hand, Expected Free Energy (EFE) and Generalized Free Energy (GFE) were proposed to extend the one-step ahead implicit action computation into an explicit policy formulation, where the agent is able to compute the best action taking into account a time horizon [16]. Initial agent implementations of these approaches needed the enumeration over every possible policy projected forward in time up to the time horizon, resulting in significant scaling limitations. As a solution, deep neural networks were also proposed to approximate the densities comprising the agent's generative model [1–6], allowing active inference to be scaled up to larger and more complex tasks.

However, despite the general theoretical formulation, current state-of-the-art dAIF, has only been successfully tested in toy problems with fully observable state spaces (Markov Decision Processes, MDP). Conversely, Deep Q-learning (DQN) approaches [7] can deal with high-dimensional inputs such as images.

Here, we propose a dAIF model[1] that extends the formulation presented in [3] to tackle problems where the state is not observable[2] (i.e. Partially Observable Markov Decision Processes, POMDP), in particular, the environment state has to be inferred directly high-dimensional from visual input. The agent's objective is to minimize its EFE into the future up to some time horizon T similarly as a receding horizon controller. We compared the performance of our proposed dAIF algorithm in the OpenAI CartPole-v1 environment against DQN. We show that the proposed approach has comparable or better performance depending on observability.

2   Deep Active Inference Model

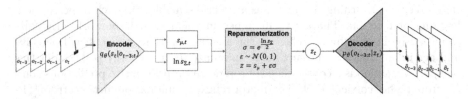

Fig. 1. Observations-state neural network architecture. The VAE encodes the visual features that are relevant to reconstruct the input images. The encoder network encodes observations to a state representation of the environment. The decoder reconstructs the input observations from this representation.

---

[1] The code is available on: https://github.com/Grottoh/Deep-Active-Inference-for-Partially-Observable-MDPs.

[2] We formulate image-based estimation and control as a POMDP—see [17] for a discussion.

We define the active inference agent's objective as optimizing its variational free energy (VFE) at time t, which can be expressed as:

$$-F_t = D_{KL}[q(s,a)\|p(o_t, s_{0:t}, a_{0:t})] \tag{1}$$
$$= -E_{q(s_t)}[\ln p(o_t|s_t)] + D_{KL}[q(s_t)\|p(s_t|s_{t-1}, a_{t-1})]$$
$$+ D_{KL}[q(a_t|s_t)\|p(a_t|s_t)] \tag{2}$$

where o_t is the observation at time t, s_t is the state of the environment, a_t is the agent's action and $E_{q(s_t)}$ is the expectation over the variational density $q(s_t)$.

We approximate the densities of Eq. 2 with deep neural networks as proposed in [1,3,4]. The first term, containing densities $q(s_t)$ and $p(o_t|s_t)$ concerns the mapping of observations to states, and vice-versa. We capture this objective with a variational autoencoder (VAE). A graphical representation of this part of the neural network architecture is depicted in Fig. 1 – see the appendix for network details.

We can use an encoder network $q_\theta(s_t|o_{t-3:t})$ with parameters θ to model $q(s_t)$, and we can use a decoder network $p_\vartheta(o_{t-3:t}|z_t)$ with parameters ϑ to model $p(o_t|s_t)$. The encoder network encodes high-dimensional input as a distribution over low-dimensional latent states, returning the sufficient statistics of a multivariate Gaussian, i.e. the mean s_μ and variance s_Σ. The decoder network consequently reconstructs the original input from reparametrized sufficient statistics z. The distribution over latent states can be used as a model of the environment in case the true state of an environment is inaccessible to the agent (i.e. in a POMDP).

The second term of Eq. 2 can be interpreted as state prediction error, which is expressed as the Kullback-Leibler (KL) divergence between the state derived at time t and the state that was predicted for time t at the previous time point. In order to compute this term the agent must, in addition to the already addressed $q(s_t)$, have a transition model $p(s_t|s_{t-1}, a_{t-1})$, which is the probability of being in a state given the previous state and action. We compute the MAP estimate with a feedforward network $\hat{s}_t = f_\phi(s_{\mu,t-1}, a_{t-1})$. To compute the state prediction error, instead of using the KL-divergence over the densities, we use the Mean-Squared-Error (MSE) between the encoded mean state s_μ and the predicted state $\hat{s}$ returned by f_ϕ

The third and final term contains the last two unaddressed densities $q(a_t|s_t)$ and $p(a_t|s_t)$. We model variational density $q(a_t|s_t)$ using a feedforward neural network $q_\xi(a_t|s_{\mu,t}, s_\Sigma)$ parameterized by ξ, which returns a distribution over actions given a multivariate Gaussian over states. Finally, we model action conditioned by the state or policy $p(a_t|s_t)$. According to the active inference literature, if an agent that minimizes the free energy does not have the prior belief that it selects policies that minimize its (expected) free energy (EFE), it would infer policies that do not minimize its free energy [16]. Therefore, we can assume that our agent expects to act as to minimize its EFE into the future. The EFE of a policy π after time t onwards can be expressed as:

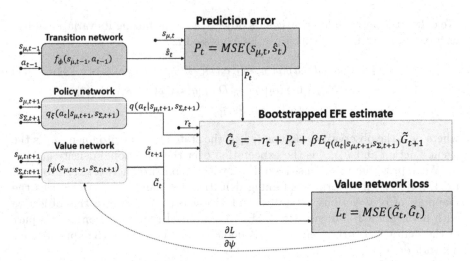

Fig. 2. Computing the gradient of the value network with the aid of a bootstrapped EFE estimate.

$$G_\pi = \sum_{\tau > t} G_{\pi,t}$$

$$G_{\pi,\tau} = \underbrace{-\ln p(o_\tau)}_{-r_\tau} + D_{KL}[q(s_\tau|\pi)\|q(s_\tau|o_\tau)] \tag{3}$$

Note that the EFE has been transformed into a RL instance by substituting the negative log-likelihood of an observation $-\ln p(o_\tau)$ (i.e. surprise) by the reward r_τ [3,18]. Since under this formulation minimizing one's EFE involves computing one's EFE for each possible policy π for potentially infinite time points τ, a tractable way to compute G_π is required. Here we estimate G_π via bootstrapping, as proposed in [3]. To this end the agent is equipped with an EFE-value (feedforward) network $f_\psi(s_{\mu,t}, s_{\Sigma,t})$ with parameters ψ, which returns an estimate $\tilde{G}_t$ that specifies an estimated EFE for each possible action. This network is trained with the aid of a bootstrapped EFE estimate $\hat{G}_t$, which consists of the free energy for the current time step, and a $\beta \in (0,1]$ discounted value net approximation of the free energy expected under $q(a|s)$ for the next time step:

$$\hat{G}_t = -r_t + D_{KL}[q(s_t)\|q(s_t|o_t)] + \beta E_{q(a_{t+1}|s_{t+1})}\tilde{G}_t \tag{4}$$

In this form the parameters of $f_\psi(s_{\mu,t}, s_{\Sigma,t})$ can be optimized through gradient descent on (see Fig. 2):

$$L_t = MSE(\tilde{G}_t, \hat{G}_t) \tag{5}$$

The distribution over actions can then at last be modelled as a precision-weighted Boltzmann distribution over our EFEs estimate [3,16]:

$$p(a_t|s_t) = \sigma(-\gamma\tilde{G}_t) \tag{6}$$

Finally, Eq. 2 is computed with the neural network density approximations as – see Fig. 3.

$$-F_t = - E_{q_\theta(s_t|o_{t-3:t})}[\ln p_\vartheta(o_{t-3:t}|z_t)]$$
$$+ MSE(s_{\mu,t}, f_\phi(s_{\mu,t-1}, a_{t-1}))$$
$$+ D_{KL}[q_\xi(a_t|s_{\mu,t}, s_{\Sigma,t})\|\sigma(-\gamma f_\psi(s_{\mu,t}, s_{\Sigma,t}))] \tag{7}$$

where $s_{\mu,t}$ and $s_{\Sigma,t}$ are encoded by $q_\theta(s_t|o_{t-3:t})$.

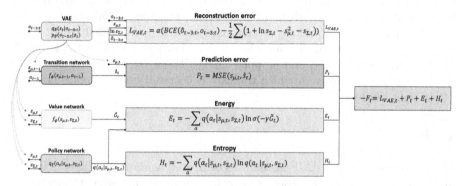

Fig. 3. Variational Free Energy computation using the approximated densities. The VAE encodes high-dimensional input as a latent state space, which is used as input to the other networks. Note that the third term of Eq. 7 $(D_{KL}[q_\xi(a_t|s_{\mu,t}, s_{\Sigma,t})\|\sigma(-\gamma f_\psi(s_{\mu,t}, s_{\Sigma,t}))])$ has been split into an energy and an entropy term (Any KL divergence can be split into an energy term and an entropy).

3 Experimental Setup

To evaluate the proposed algorithm we used the OpenAI Gym's CartPole-v1, as depicted in Fig. 4. In the CartPole-v1 environment, a pole is attached to a cart that moves along a track. The pole is initially upright, and the agent's objective is to keep the pole from tilting too far to one side or the other by increasing or decreasing the cart's velocity. Additionally, the position of the cart must remain within certain bound. An episode of the task terminates when the agent fails

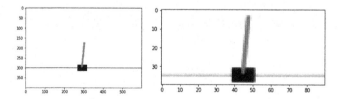

Fig. 4. Cartpole-v1 benchmark (left) and cropped visual input used (right).

either of these objectives, or when it has survived for 500 time steps. Each time step the agent receives a reward of 1.

The CartPole state consists of four continuous values: the cart position, the cart velocity, the pole angle and the velocity of the pole at its tip. Each run the state values are initialized at random within a small margin to ensure variability between runs. The agent can exact influence on the next state through two discrete actions, by pushing the cart to the left, or by pushing it to the right.

Tests were conducted in two scenarios: 1) an MDP scenario in which the agent has direct access to the state of the environment, and 2) a POMDP scenario in which the agent does not have direct access the environment state, and instead receives pixel value from which it must derive meaningful hidden states. By default, rendering the CartPole-v1 environment returns a $3 \times 400 \times 600$ (color, height, width) array of pixel values. In our experiments we provide the POMDP agents with a downscaled and cropped image. There the agents receive a $3 \times 37 \times 85$ pixel value array in which the cart is centered until it comes near the left or right border.

4   Results

The performance of our dAIF agents was compared against DQN agents for the MDP and the POMDP scenarios, and against an agent that selects it actions at random. Each agent was equipped with a memory buffer and a target network [19]. The memory buffer stores transitions from which the agent can sample random batches on which to perform batch gradient descent. The target network is a copy of the value network of which the weights are not updated directly through gradient descent, but are instead updated periodically with the weights of the value network. In between updates this provides the agent with fixed EFE-value or Q-value targets, such that the value network does not have to chase a constantly moving objective.

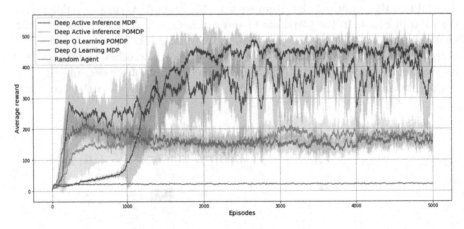

Fig. 5. Average reward comparison for the CartPole-v1 problem.

The VAE of the POMDP dAIF agent is pre-trained to deconstruct input images into a distribution over latent states and to subsequently reconstruct them as accurately as possible.

Figure 5 shows the mean and standard deviation of the moving average reward (MAR) over all runs for the five algorithms at each episode. Each agent performed 10 runs of 5000 episodes. The moving average reward for an episode e is calculated using an smoothing average:

$$MAR_e = 0.1CR_e + 0.9MAR_{e-1} \tag{8}$$

Where CR_e is the cumulative reward of episode e and MAR_{e-1} is the MAR of the previous episode.

The dAIF MDP agent results closely resemble those presented in [3] and outperforms the DQN MDP agent by a significant margin. Further, the standard deviation shadings show that the dAIF MDP is agent is more consistent between runs than the DQN agent. The POMDP agents are both demonstrated to be capable of learning successful policies, attaining comparable performance.

We have exploited probabilistic model based control through a VAE that encodes the state. On the one hand, this allows the tracking of an internal state which can be used for a range of purposes, such the planning of rewarding policies and the forming of expectations about the future. On the other hand, it makes every part of the algorithm dependent on the proper encoding of the latent space, conversely to the DQN that did not require a state representation to achieve the same performance. However, we expect our approach to improve relative to DQN in more complex environments where the world state encoding can play a more relevant role.

5 Conclusion

We described a dAIF model that tackles partially observable state problems, i.e., it learns the policy from high-dimensional inputs, such as images. Results show that in the MDP case the dAIF agent outperforms the DQN agent, and performs more consistently between runs. Both agents were also shown to be capable of learning (less) successful policies in the POMDP version, where the performance between dAIF and DQN models was found to be comparable. Further work will focus on validating the model on a broader range of more complex problems.

Appendix

Deep Q Agent MDP	
Networks & params.	Description
N_s	Number of states.
N_a	Number of actions.
Q-value network	Fully connected network using an Adam optimizer with a learning rate of 10^{-3}, of the form: $N_s \times 64 \times N_a$.
γ	Discount factor set to 0.98
Memory size	Maximum amount of transition that can be stored in the memory buffer: 65,536
Mini-batch size	32
Target network freeze period	The amount of time steps the target network's parameters are frozen, until they are updated with the parameters of the value network: 25

Deep Q Agent POMDP	
Networks & params.	Description
N_a	Number of actions.
Q-value network	Consists of three 3D convolutional layers (each followed by batch normalization and a rectified linear unit) with $5 \times 5 \times 1$ kernels and $2 \times 2 \times 1$ strides with respectively 3, 16 and 32 input channels, ending with 32 output channels. The output is fed to a 2048×1024 fully connected layer which leads to a $1024 \times N_a$ fully connected layer. Uses an Adam optimizer with the learning rate set to 10^{-5}.
γ	Discount factor set to 0.99
Memory size	Maximum amount of transition that can be stored in the memory buffer: 65,536
Mini-batch size	32
Target network freeze period	The amount of time steps the target network's parameters are frozen, until they are updated with the parameters of the value network: 25

Deep Active Inference Agent MDP	
Networks & params.	Description
N_s	Number of states.
N_a	Number of actions.
Transition network	Fully connected network using an Adam optimizer with a learning rate of 10^{-3}, of the form: $(N_s + 1) \times 64 \times N_s$.
Policy network	Fully connected network using an Adam optimizer with a learning rate of 10^{-3}, of the form: $N_s \times 64 \times N_a$, a softmax function is applied to the output.
EFE-value network	Fully connected network using an Adam optimizer with a learning rate of 10^{-4}, of the form: $N_s \times 64 \times N_a$.
γ	Precision parameter set to 1.0
β	Discount factor set to 0.99
Memory size	Maximum amount of transition that can be stored in the memory buffer: 65,536
Mini-batch size	32
Target network freeze period	The amount of time steps the target network's parameters are frozen, until they are updated with the parameters of the value network: 25

Deep Active Inference Agent POMDP	
Networks & params.	Description
N_l	Size of the VAE latent space, here set to 32.
N_a	Number of actions.
Encoder-network $q_\theta(s_t\|o_{t-3:t})$	Consists of three 3D convolutional layers (each followed by batch normalization and a rectified linear unit) with $5 \times 5 \times 1$ kernels and $2 \times 2 \times 1$ strides with respectively 3, 16 and 32 input channels, ending with 32 output channels. The output is fed to a 2048×1024 fully connected layer which splits to two additional $1024 \times N_l$ fully connected layers. Uses an Adam optimizer with the learning rate set to 10^{-5}.
Decoder-network $p_\vartheta(o_{t-3:t}\|z_t)$	Consists of a $N_l \times 1024$ fully connected layer leading to a 1024×2048 fully connected layer leading to three 3D transposed convolutional layers (each followed by batch normalization and a rectified linear unit) with $5 \times 5 \times 1$ kernels and $2 \times 2 \times 1$ strides with respectively 32, 16 and 3 input channels, ending with 3 output channels. Uses an Adam optimizer with the learning rate set to 10^{-5}.
Transition-network $f_\phi(s_{\mu,t}, a_t)$	Fully connected network using an Adam optimizer with a learning rate of 10^{-3}, of the form: $(2N_l + 1) \times 64 \times N_l$.
Policy-network $q_\xi(s_{\mu,t}, s_{\Sigma,t})$	Fully connected network using an Adam optimizer with a learning rate of 10^{-3}, of the form: $2N_l \times 64 \times N_a$, a softmax function is applied to the output.
EFE-value-network $f_\psi(s_{\mu,t}, s_{\Sigma,t})$	Fully connected network using an Adam optimizer with a learning rate of 10^{-4}, of the form: $2N_l \times 64 \times N_a$.
γ	Precision parameter set to 12.0
β	Discount factor set to 0.99
α	A constant that is multiplied with the VAE loss to take it to the same scale as the rest of the VFE terms, set to 4×10^{-5}
Memory size	Maximum amount of transition that can be stored in the memory buffer: 65,536
Mini-batch size	32
Target network freeze period	The amount of time steps the target network's parameters are frozen, until they are updated with the parameters of the value network: 25

References

1. Ueltzhöffer, K.: Deep active inference. Biol. Cybern. **112**(6), 547–573 (2018). https://doi.org/10.1007/s00422-018-0785-7
2. Sancaktar, C., van Gerven, M., Lanillos, P.: End-to-end pixel-based deep active inference for body perception and action. arXiv preprint arXiv:2001.05847 (2019)
3. Millidge, B.: Deep active inference as variational policy gradients. J. Math. Psychol. **96**, 102348 (2020). https://doi.org/10.1016/j.jmp.2020.102348
4. Tschantz, A., Baltieri, M., Seth, A.K., Buckley, C.L.: Scaling active inference. arXiv Prepr. arXiv:1911.10601v1 (2019)
5. Çatal, O., Wauthier, S., Verbelen, T., Boom, C.D., Dhoedt, B.: Deep active inference for autonomous robot navigation. arXiv Prepr. arXiv:2003.03220v1 (2020)
6. Fountas, Z., Sajid, N., Mediano, P.A.M., Friston, K.: Deep active inference agents using monte-carlo methods. arXiv Prepr. arXiv:2006.04176v1 (2020)
7. Mnih, V., et al.: Playing atari with deep reinforcement learning. arXiv Prepr. arXiv:1312.5602v1 (2013)
8. Arulkumaran, K., Deisenroth, M.P., Brundage, M., Bharath, A.A.: Deep reinforcement learning: a brief survey. IEEE Signal Process. Mag. **34**(6), 26–38 (2017)
9. Friston, K.J., Daunizeau, J., Kilner, J., Kiebel, S.J.: Action and behavior: a free-energy formulation. Biol. Cybern. **102**, 227–260 (2010). https://doi.org/10.1007/s00422-010-0364-z
10. Friston, K.J.: The free-energy principle: a unified brain theory? Nature **11**, 127–138 (2010). https://doi.org/10.1038/nrn2787
11. Adams, R.A., Shipp, S., Friston, K.J.: Predictions not commands: active inference in the motor system. Brain Struct. Funct. **218**(3), 611–643 (2012). https://doi.org/10.1007/s00429-012-0475-5
12. Lanillos, P., Pages, J., Cheng, G.: Robot self/other distinction: active inference meets neural networks learning in a mirror. In: Proceedings of the 24th European Conference on Artificial Intelligence (ECAI), pp. 2410–2416 (2020). https://doi.org/10.3233/FAIA200372
13. Rood, T., van Gerven, M., Lanillos, P.: A deep active inference model of the rubber-hand illusion. arXiv Prepr. arXiv:2008.07408 (2020)
14. Oliver, G., Lanillos, P., Cheng, G.: Active inference body perception and action for humanoid robots. arXiv preprint arXiv:1906.03022 (2019)
15. Friston, K., Trujillo-Barreto, N., Daunizeau, J.: Dem: a variational treatment of dynamic systems. NeuroImage **41**(3), 849–885 (2008). https://doi.org/10.1016/j.neuroimage.2008.02.054
16. Parr, T., Friston, K.J.: Generalised free energy and active inference. Biol. Cybern. **113**, 495–513 (2019). https://doi.org/10.1007/s00422-019-00805-w
17. Hausknecht, M., Stone, P.: Deep recurrent q-learning for partially observable MDPs. In: AAAI Fall Symposium on Sequential Decision Making for Intelligent Agents (AAAI-SDMIA15), November 2015
18. Friston, K.J., Samothrakis, S., Montague, R.: Active inference and agency: optimal control without cost functions. Biol. Cybern. **106**, 523–541 (2012). https://doi.org/10.1007/s00422-012-0512-8
19. Mnih, V., et al.: Human-level control through deep reinforcement learning. Nature **518**, 529–533 (2015). https://doi.org/10.1038/nature14236

Sleep: Model Reduction in Deep Active Inference

Samuel T. Wauthier[✉], Ozan Çatal, Cedric De Boom, Tim Verbelen,
and Bart Dhoedt

IDLab, Department of Information Technology at Ghent University – imec,
Technologiepark-Zwijnaarde 126, 9052 Ghent, Belgium
{samuel.wauthier,ozan.catal,cedric.deboom,tim.verbelen,
bart.dhoedt}@ugent.be

Abstract. Sleep is one of the most important states of the human mind
and body. Sleep has various functions, such as restoration, both physically and mentally, and memory processing. The theory of active inference frames sleep as the minimization of complexity and free energy in
the absence of sensory information. In this paper, we propose a method
for model reduction of neural networks that implement the active inference framework. The proposed method suggests initializing the network
with a high latent space dimensionality and pruning dimensions subsequently. We show that reduction of latent space dimensionality decreases
complexity without increasing free energy.

Keywords: Active inference · Model reduction · Sleep

1 Introduction

Sleep is a phenomenon that occurs in most animals [10]. It is a topic of intensive
research as it has been shown to be important for both the mind [12,18] and
the body [14]. In particular, sleep and learning have been connected in many
hypotheses [3,17], as well as mental health [4] and memory [20].

Active inference is a theory of behaviour and learning that originated in neuroscience [8]. The basic assumption is that intelligent agents attempt to minimize
their variational free energy. Variational free energy—named for its counterpart
in statistical physics i.e. Helmholtz free energy—is also known as the evidence
lower bound (ELBO) in variational Bayesian methods.

Since its conception, active inference has been explored in multiple subfields
of neuroscience and biology [5,6,11] and eventually found its way into the field
of computer science [2,15,19]. In particular, Ueltzhöffer [19] and Çatal *et al.* [2]
have made developments in *deep active inference*, i.e. the use of deep neural
networks to implement active inference.

Recent work [7,9,13] has pointed out the relation between the function of
removing redundant connections during sleep and Bayesian model reduction
(BMR) in active inference, i.e. complexity minimization through elimination

© Springer Nature Switzerland AG 2020
T. Verbelen et al. (Eds.): IWAI 2020, CCIS 1326, pp. 72–83, 2020.
https://doi.org/10.1007/978-3-030-64919-7_9

of redundant parameters. In this work, we propose a method for reducing complexity in the deep active inference framework. We evaluate the method through simulation experiments.

2 Deep Active Inference

Currently, using deep neural networks in active inference to learn state spaces, in addition to policy and posterior, is becoming increasingly popular, which contrasts with active inference on discrete state spaces as described in [9]. In this approach, the dimensionality of the state space is a hyperparameter, i.e. it must be specified before training and cannot change along the way. Here, we briefly introduce the method provided by Çatal et al. [2].

Assuming the policy π may be broken up into a sequence of actions $\mathbf{a}_t$ and the current state depends on the previous action instead of the policy, a generative model with observations $\mathbf{o}_t$ and states $\mathbf{s}_t$ is defined as

$$P(\tilde{\mathbf{o}}, \tilde{\mathbf{s}}, \tilde{\mathbf{a}}) = P(\mathbf{s}_0)P(\tilde{\mathbf{a}}) \prod_{t=1}^{T} P(\mathbf{o}_t|\mathbf{s}_t)P(\mathbf{s}_t|\mathbf{s}_{t-1}, \mathbf{a}_{t-1}), \tag{1}$$

where $\tilde{\mathbf{x}} = (\mathbf{x}_0, \mathbf{x}_1, \mathbf{x}_2, \ldots, \mathbf{x}_T)$.

Deep neural networks are used to parameterize the prior, likelihood and approximate posterior distributions: $p_\theta(\mathbf{s}_t|\mathbf{s}_{t-1}, \mathbf{a}_{t-1})$, $p_\phi(\mathbf{o}_t|\mathbf{s}_t)$ and $q_\xi(\mathbf{s}_t|\mathbf{s}_{t-1}, \mathbf{a}_{t-1}, \mathbf{o}_t)$, respectively. With this, minimization of free energy consists of minimizing the loss function

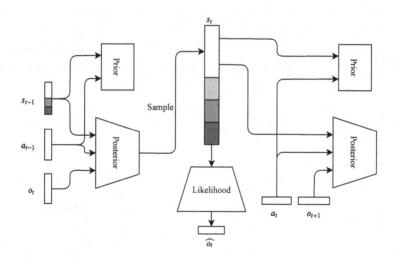

Fig. 1. Information flow of neural networks. The posterior network takes in previous state and action, and current observation. The prior network takes in previous state and action. The likelihood network takes in current state. The state s_t illustrates that dimensions may be pruned, if they are unused.

$$L(\theta, \phi, \xi; \mathbf{o}_t, \mathbf{s}_{t-1}, \mathbf{a}_{t-1}) =$$
$$D_{\mathrm{KL}}(q_\xi(\mathbf{s}_t|\mathbf{s}_{t-1}, \mathbf{a}_{t-1}, \mathbf{o}_t)||p_\theta(\mathbf{s}_t|\mathbf{s}_{t-1}, \mathbf{a}_{t-1})) - \log p_\phi(\mathbf{o}_t|\mathbf{s}_t). \qquad (2)$$

Prior, likelihood and posterior distributions are chosen to be multivariate normal distributions. As opposed to the standard VAE, optimization is done over sequences in time. Additionally, empirical priors are learned, instead of using fixed priors. A chart on the information flow can be found in Fig. 1.

3   Latent Space Dimensionality Reduction and Sleep

The size of the latent space vector $\mathbf{s}$ is an important hyperparameter. On the one hand, this must be large enough to explain observations in the generative model. On the other hand, it must be kept minimal to reduce complexity as to minimize the required resources, such as memory and power (both computational and electrical). In general, one does not know the optimal size of $\mathbf{s}$. A typical way of finding a well-performing value is a hyperparameter sweep. Parameter sweeps, however, are resource intensive and require many unnecessary training runs. Therefore, we propose a method for dimensionality reduction in the deep active inference framework.

The basic idea is to prune dimensions in the latent space vector $\mathbf{s}$. A popular method for inspecting informative dimensions of a vector space is singular value decomposition (SVD). This technique is used to factorize an $m \times n$ matrix $\boldsymbol{A}$ into three matrices $\boldsymbol{USV}^*$, where a common geometrical interpretation is that the decomposition gives 2 rotation matrices $\boldsymbol{U}$ and $\boldsymbol{V}^*$, and a scaling matrix $\boldsymbol{S}$.

Algorithm 1 lines out the method in the form of pseudo-code. Let n be the dimensionality of the latent space. We sample a latent space vector from m

---

Algorithm 1: Sleep

---

input : A trained model model with dimensionality n
The number of repetitions N and number of sequences m
A threshold α

output: The new latent space dimensionality ν

```
while i < N do
    A ← []                                    // make a matrix
    while j < m do
        a ← GenerateSequence(model)           // generate a new sequence
        v ← Sample(a)                         // sample a latent space vector
        A ← [A, v]              // insert vector as a new column in matrix
        j ← j + 1
    S ← SVD(A)                                // apply SVD to matrix
    c_i ← #(S_kk > α) for 0 < k < n           // count sv's over threshold
    c ← [c, c_i]                  // add number to list of outcomes
    i ← i + 1
ν ← Avg(c)                                    // average over all outcomes
```

---

different sequences to construct the column vectors of a matrix A. The column space of A, denoted $\mathcal{C}(A)$, forms a subspace of the latent space. Applying SVD to A gives the scaling matrix S. The values on the diagonal of S are the singular values of A, and suggest a size for the dimensions of $\mathcal{C}(A)$ after rotation with U. Dimensions with small singular values are assumed to be unused. To this end, we define a threshold α for which dimensions corresponding to singular values smaller than α can be pruned. We repeat this procedure N times—by generating m new sequences each time—and average the number of pruned dimensions, in order to obtain a relatively robust outcome.

It is important to stress, here, that SVD does not allow one to find *which* dimensions can be pruned. Instead, it is used to converge to the optimal *number* of dimensions. SVD provides the size of dimensions of the column space of A, i.e. $\mathcal{C}(A)$, described in a basis of the latent space after a rotation with U. The actual basis vectors are a linear combination of the rotated basis vectors. In other words, having a zero dimension in rotated latent space, does not necessarily mean there is one in latent space. However, it does indicate that it is possible to reduce dimensionality by choosing a different rotation, since it shows that there is an orientation of the basis vectors which requires less dimensions to describe the column space. Returning to the model, by retraining with a lower dimensionality n, we essentially force the model to learn the latent space with a different orientation which requires less dimensions.

We have dubbed the method *sleep*, since it replicates synapse pruning, as well as Bayesian model reduction. From an active inference perspective, the proposed method is analogous to BMR in that it considers a generative model with a large number of latent factors and optimizes this number post hoc [16]. In other words, both the goals of the proposed method and BMR are to consider alternative models which may give simpler explanations for the same observations. That said, in both cases, the balance between accuracy and complexity is crucial, i.e. accuracy should not suffer due to simplicity. Indeed, the measure for this trade-off is free energy.

Since latent space in deep active inference is learned using deep neural networks, there is no guarantee that each latent space dimension represents an individual feature. Without knowing what is contained in latent space, it is not possible to target specific parameters to turn off as in BMR. Because of this, Algorithm 1 must be succeeded by retraining to obtain a reduced model. In this sense, the earlier analogy is incomplete, since BMR allows one to obtain the reduced model parameters from the full model, i.e. it allows one to find *which* dimensions can be pruned.

In the end, the purpose of the sleep method is to reduce complexity whenever an application (e.g.. a robot running a deep active inference implementation) has downtime. The overall sequence of events, then, proceeds as follows. Start with a large value for the latent space dimensionality and train the model. Deploy the model on the application. Each time there is downtime in the application (e.g.. the robot is charging), reduce the model by sleeping and retraining. Continue this pattern of sleeping and retraining until the model cannot reduce any further.

4   Experimental Setup

Experiments were performed using two environments from the OpenAI Gym [1]. The first experiment employs a modified version of the MountainCar environment, where noise is added to the observation and only the position can be observed. The goal of this environment is to drive up a steep mountain using an underpowered car that starts in a valley. The car is underpowered in the sense that it cannot produce enough force to go against gravity and drive up the mountain in one go. It must first build up enough momentum by driving up the side(s) of the valley. In this experiment, we know upfront that the model only needs 2 dimensions in latent space: position and velocity. Details about the neural networks used for this experiment can be found in Appendix B.1.

The second experiment employs the CarRacing environment. The goal of this environment is to stay in the middle of a race track using a race car. The car and track are viewed from a top-down perspective. The car must steer left and right to stay on track. Compared to the MountainCar, the CarRacing environment utilizes more complicated dynamics and produces higher dimensional observations. Examples of the environments can be found in Appendix A. Details about the neural networks used for this experiment can be found in Appendix B.2.

5   Results

Figure 2 shows the evolution of the free energy of MountainCar during training with a fixed number of latent space dimensions (see Appendix C.1 for a similar figure for CarRacing). It suggests that free energy decreases as more state space dimensions are added. However, it also shows that free energy does not visibly decrease beyond a certain number of dimensions. For the MountainCar, we see that the free energy does not decrease for more than 2 dimensions, while for the CarRacing (Appendix C.1), we see that the free energy does not decrease for more than 4 dimensions. In essence, there appears to be a critical value of

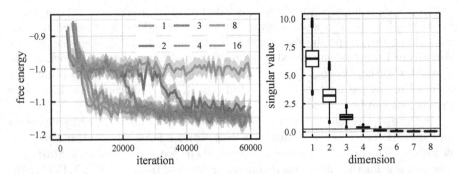

Fig. 2. (Left) Free energy during training of MountainCar for different state space sizes. Curves show smoothed data (LOESS, span 0.02) with 95% standard error bands. (Right) Boxplot of singular values while sleeping at 8 latent space dimensions ($N = 10^4$). (Color figure online)

the latent space size. For latent spaces larger than this critical value, the free energy does not reduce. This critical value corresponds to the optimal value for the dimensionality with respect to the accuracy/complexity trade-off.

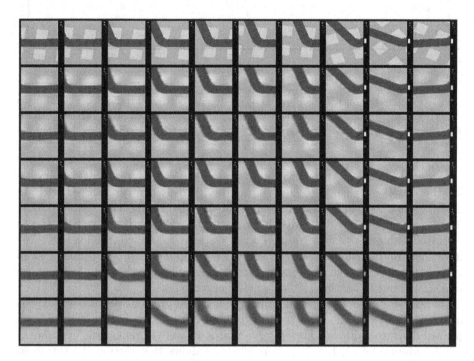

Fig. 3. Reconstructions of CarRacing track over time with different latent space dimensions. From top to bottom: ground truth, 32, 16, 8, 4, 2, 1.

Figure 3 demonstrates how latent space dimensionality affects reconstruction and how too few dimensions can lead to aspects of the environment not being learned. It shows reconstructions of the CarRacing track for latent space dimensions of 32, 16, 8, 4, 2 and 1 (top to bottom with ground truth in the top sequence). Note how the curvature of the track is not accurately reconstructed through 1 dimension, especially at early time steps. Also, 2 dimensions still seem to lack accuracy (see curvature in second time step). Furthermore, note how the feedback bar is incorrectly encoded by dimensions lower than 4.

Figure 2 also shows a boxplot for the singular values obtained for the MountainCar with 8 latent space dimensions for $N = 10^4$ iterations (plots for different latent space dimensions can be found in Appendix C.2). The red line shows a threshold $\alpha = 0.25$. The figure suggests that there is a difference in sizes in the latent space dimensions. Indeed, the first four singular values are on average larger than α, while the remaining values are on average smaller than α. This indicates that certain dimensions are very small, therefore, contain less information, and may be pruned subsequently.

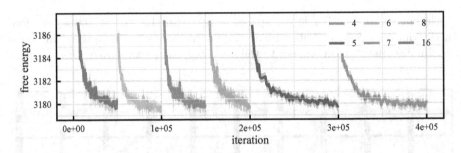

Fig. 4. Free energy over 7 sleep cycles of CarRacer. Setting threshold $\alpha = 0.25$ gives the reduction: $16 \rightarrow 8 \rightarrow 7 \rightarrow 6 \rightarrow 5 \rightarrow 4$, after which it cannot reduce further. Curves show smoothed data (LOESS, span 0.02) with 95% standard error bands.

Figure 4 illustrates the algorithm put into practice with different sleep cycles for the CarRacer with threshold set at $\alpha = 0.25$, where we started with 16 latent space dimensions. In this example, we initiated sleep every 5×10^4 training iterations and checked if dimensionality could be reduced. If so, we pruned and restarted training with lower dimensionality, else we continued training for 5×10^4 iterations, until reduction was possible. We stopped the process after 7 sleep cycles. As expected, the sleep sequence manages to reduce the complexity of the model, without impacting the free energy negatively.

When compared to Fig. 8 in Appendix C.2, the previous result is exactly as expected. Following the steps described there, the state space can effectively be pruned down to 4 dimensions. Observe that if we were to repeat the experiment for the MountainCar, Fig. 7 in Appendix C.2 shows that setting the threshold at $\alpha = 0.25$ would return a state space dimensionality of 2.

6   Conclusion

Our results show that it is possible to train a deep active inference model by setting a large number of latent space dimensions and subsequently sleeping until minimal complexity is reached. However, the method proposed in this paper is not optimal. A few caveats remain. First of all, the current method requires retraining. After applying SVD, the entire model must be retrained from scratch. Second of all, there exist limitations to SVD. For instance, SVD does not take into account nonlinear transformations. Therefore, relations between different dimensions may remain and the optimal dimensionality may never be reached.

In future work, we will investigate the effects of sleeping at regular intervals during training. For example, we may sleep after every 10^4 time steps to check if we can already reduce the latent space. Another option we will investigate, is to prune both unnecessary dimensions and weights. This way, we may be able to maintain the trained neural network, while reducing complexity. In addition, we want to experiment with different methods for dimensionality reduction, such as nonlinear methods. Another option to be explored is to learn and set unused dimensions to 0 during training.

Acknowledgments. This research received funding from the Flemish Government under the "Onderzoeksprogramma Artificiële Intelligentie (AI) Vlaanderen" programme.

Appendix A OpenAI Gym Examples

Figure 5 shows snapshots of the MountainCar and CarRacing environments from the OpenAI Gym [1]. Note that observations in the MountainCar environment consist of position and velocity values, while CarRacing provides RGB pixels.

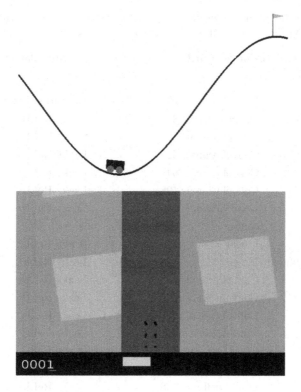

Fig. 5. (Top) Example of MountainCar environment [1]. (Bottom) Example of Car-Racing environment [1].

Appendix B Neural Network Definitions

Appendix B.1 Mountain Car

Table 1 shows the neural architecture of the network used in the MountainCar experiments.

Table 1. Specifications of the MountainCar neural network with s latent space dimensions.

	Layer	Neurons/Filters	Activation function
Posterior	Linear	20	Leaky ReLU
	Linear	$2 \times s$	Leaku ReLU
Likelihood	Linear	20	Leaky ReLU
	Linear	2	Leaky ReLU
Prior	Linear	20	Leaky ReLU
	Linear	$2 \times s$	Leaky ReLU

Appendix B.2 Car Racing

Table 2 shows the neural architecture of the network used in the CarRacing experiments. All filters have 3×3 kernel, as well as stride and padding of 1.

Table 2. Specifications of the CarRacing neural network with s latent space dimensions.

	Layer	Neurons/Filters	Activation function
Posterior	Convolutional	8	Leaky ReLU
	Convolutional	16	Leaky ReLU
	Convolutional	32	Leaky ReLU
	Convolutional	64	Leaky ReLU
	Convolutional	128	Leaky ReLU
	Convolutional	256	Leaky ReLU
	concat	N/A	N/A
	Linear	$2 \times s$	Leaku ReLU
Likelihood	Linear	$128 \times 2 \times 9$	Leaky ReLU
	Convolutional	128	Leaky ReLU
	Convolutional	64	Leaky ReLU
	Convolutional	32	Leaky ReLU
	Convolutional	16	Leaky ReLU
	Convolutional	8	Leaky ReLU
	Convolutional	3	Leaky ReLU
Prior	LSTM cell	128	Leaky ReLU
	Linear	$2 \times s$	Softplus

Appendix C Additional Plots

Appendix C.1 Free Energy During Training

Figure 6 shows the evolution of the free energy during training for CarRacing similar to the left plot in Fig. 2. Note how the free energy does not visibly decrease when using more than 4 dimensions.

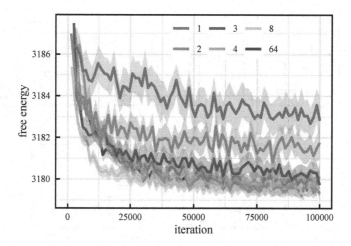

Fig. 6. Free energy during training of CarRacer for different state space sizes. Curves show smoothed data (LOESS, span 0.02) with 95% standard error bands.

Appendix C.2 Boxplots for Different Latent Space Dimensions

Figure 7 shows boxplots for the singular values obtained for the MountainCar with different latent space dimensions for 10^4 iterations, while Fig. 8 shows the same for the CarRacing. The red line in each plot indicates the threshold $\alpha = 0.25$.

One can do the following mental exercise. Choose a boxplot and count the amount of dimensions that are above threshold on average. This number will be the new dimensionality. Go to the boxplot for that dimensionality and, again, count the amount of dimensions. Repeat this until the dimensionality does not

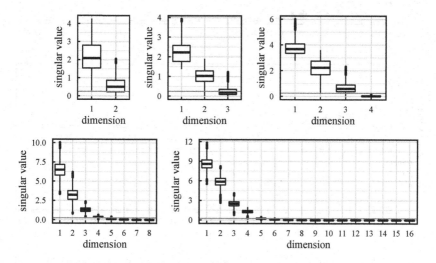

Fig. 7. Boxplots of singular values while sleeping at different latent space dimensions for the MountainCar ($N = 10^4$).

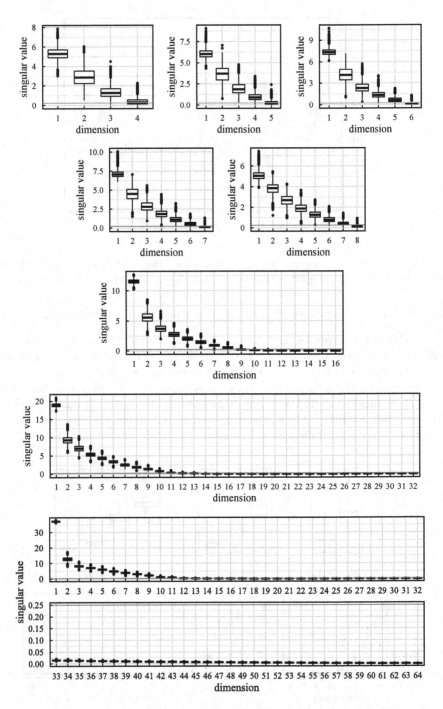

Fig. 8. Boxplots of singular values while sleeping at different latent space dimensions for the CarRacing ($N = 10^4$). (Color figure online)

reduce further. Using this process, we can see that the MountainCar will not reduce below 2 and the CarRacing will not reduce below 4.

References

1. Brockman, G., et al.: Openai gym (2016)
2. Çatal, O., Nauta, J., Verbelen, T., Simoens, P., Dhoedt, B.: Bayesian policy selection using active inference, pp. 1–9 , April 2019. http://arxiv.org/abs/1904.08149
3. Fattinger, S., et al.: Deep sleep maintains learning efficiency of the human brain. Nat. Commun. **8**(1), 15405 (2017)
4. Freeman, D., et al.: The effects of improving sleep on mental health (oasis): a randomised controlled trial with mediation analysis. Lancet Psychiatry **4**(10), 749–758 (2017)
5. Friston, K., FitzGerald, T., Rigoli, F., Schwartenbeck, P., O'Doherty, J., Pezzulo, G.: Active inference and learning. Neurosci. Biobehav. Rev. **68**, 862–879 (2016)
6. Friston, K., Mattout, J., Kilner, J.: Action understanding and active inference. Biol. Cybern. **104**(1–2), 137–160 (2011). https://doi.org/10.1007/s00422-011-0424-z
7. Friston, K., Parr, T., Zeidman, P.: Bayesian model reduction, pp. 1–32 (2018). http://arxiv.org/abs/1805.07092
8. Friston, K.J., Daunizeau, J., Kiebel, S.J.: Reinforcement learning or active inference? PLoS ONE **4**(7), e6421 (2009)
9. Friston, K.J., Lin, M., Frith, C.D., Pezzulo, G., Hobson, J.A., Ondobaka, S.: Active inference, curiosity and insight. Neural Comput. **29**(10), 2633–2683 (2017)
10. Joiner, W.J.: Unraveling the evolutionary determinants of sleep. Curr. Biol. **26**(20), R1073–R1087 (2016)
11. Kirchhoff, M., Parr, T., Palacios, E., Friston, K., Kiverstein, J.: The Markov blankets of life: autonomy, active inference and the free energy principle. J. R. Soc. Interface **15**(138), 20170792 (2018)
12. Krause, A.J., et al.: The sleep-deprived human brain. Nat. Rev. Neurosci. **18**(7), 404–418 (2017)
13. Li, W., Ma, L., Yang, G., Gan, W.B.: Rem sleep selectively prunes and maintains new synapses in development and learning. Nat. Neurosci **20**(3), 427–437 (2017)
14. Newman, A.B., et al.: Relation of sleep-disordered breathing to cardiovascular disease risk factors: the sleep heart health study. Am. J. Epidemiol. **154**(1), 50–59 (2001)
15. Oliver, G., Lanillos, P., Cheng, G.: Active inference body perception and action for humanoid robots (2019) http://arxiv.org/abs/1906.03022
16. Smith, R., Schwartenbeck, P., Parr, T., Friston, K.J.: An active inference approach to modeling structure learning: concept learning as an example case. Front. Comput. Neurosci. **14**(May), 1–24 (2020). https://doi.org/10.3389/fncom.2020.00041
17. Stickgold, R., Hobson, J.A., Fosse, R., Fosse, M.: Sleep, learning, and dreams: Off-line memory reprocessing. Science **294**(5544), 1052–1057 (2001)
18. Tsuno, N., Besset, A., Ritchie, K.: Sleep and depression. J. Clin. Psychiatry **66**(10), 1254–1269 (2005)
19. Ueltzhöffer, K.: Deep active inference. Biol. Cyber. **112**(6), 547–573 (2018). https://doi.org/10.1007/s00422-018-0785-7
20. Walker, M.P., Stickgold, R.: Sleep, memory, and plasticity. Annu. Rev. Psychol. **57**(1), 139–166 (2006)

A Deep Active Inference Model
of the Rubber-Hand Illusion

Thomas Rood[✉], Marcel van Gerven, and Pablo Lanillos

Department of Artificial Intelligence, Donders Institute for Brain,
Cognition and Behaviour, Montessorilaan 3, 6525 HR Nijmegen, The Netherlands
t.rood@student.ru.nl, p.lanillos@donders.ru.nl

Abstract. Understanding how perception and action deal with sensorimotor conflicts, such as the rubber-hand illusion (RHI), is essential to understand how the body adapts to uncertain situations. Recent results in humans have shown that the RHI not only produces a change in the perceived arm location, but also causes involuntary forces. Here, we describe a deep active inference agent in a virtual environment, which we subjected to the RHI, that is able to account for these results. We show that our model, which deals with visual high-dimensional inputs, produces similar perceptual and force patterns to those found in humans.

Keywords: Active inference · Rubber-hand Illusion · Free-energy optimization · Deep learning

1 Introduction

The complex mechanisms underlying perception and action that allow seamless interaction with the environment are largely occluded from our consciousness. To interact with the environment in a meaningful way, the brain must integrate noisy sensory information from multiple modalities into a coherent world model, from which to generate and continuously update an appropriate action [13]. Especially, how the brain-body deals with sensorimotor conflicts [8,16], e.g., conflicting information from different senses, is an essential question for both cognitive science and artificial intelligence. Adaptation to unobserved events and changes in the body and the environment during interaction is a key characteristic of body intelligence that machines still fail at.

The rubber-hand illusion (RHI) [2] is a well-known experimental paradigm from cognitive science that allows the investigation of body perception under conflicting information in a controlled setup. During the experiment, human participants cannot see their own hand but rather perceive an artificial hand placed in a different location (e.g. 15 cm from their current hand). After a minute of visuo-tactile stimulation [10], the perceived location of the real hand drifts towards the location of the artificial arm and suddenly the new hand becomes part of their own.

T. Verbelen et al. (Eds.): IWAI 2020, CCIS 1326, pp. 84–91, 2020.
https://doi.org/10.1007/978-3-030-64919-7_10

We can find some RHI modelling attempts in the literature; see [12] for an overview until 2015. In [18], a Bayesian causal inference model was proposed to estimate the perceived hand position after stimulation. In [8] a model inspired by the free-energy principle [5] was used to synthetically test the RHI in a robot. The perceptual drift (mislocalization of the hand) was compared to that of humans observations.

Recent experiments have shown that humans also generate meaningful force patterns towards the artificial hand during the RHI [1,16], adding the action dimension to this paradigm. We hypothesise that the strong interdependence between perception and action can be accounted for by mechanisms underlying active inference [7].

In this work, we propose a *deep active inference* model of the RHI, based on [14,17,19], where an artificial agent directly operates in a 3D virtual reality (VR) environment[1]. Our model 1) is able to produce similar perceptual and active patterns to human observations during the RHI and 2) provides a scalable approach for further research on body perception and active inference, as it deals with high-dimensional inputs such as visual images originated from the 3D environment.

2   Deep Active Inference Model

We formalise body perception and action as an inference problem [3,7,11,17]. The unobserved body state is inferred from the senses (observations) while taking into account its state prior information. To this end, the agent makes use of two sensory modalities. The visual input s_v is described by a pixel matrix (image) and the proprioceptive information s_p represents the angle of every joint of the arm – See Fig. 1a.

Computation of the body state is performed by optimizing the variational free-energy bound [7,17]. Under the mean-field and Laplace approximations and defining μ as the brain variables that encode the variational density that approximates the body state distribution and defining a as the action exerted by the agent, perception and action are driven by the following system of differential equations (see [4,6,19] for a derivation):

$$\dot{\mu} = -\partial_\mu F = -\partial_\mu e_p^T \Sigma_p^{-1} e_p - \partial_\mu e_v^T \Sigma_v^{-1} e_v - \partial_\mu e_f^T \Sigma_\mu^{-1} e_f \tag{1}$$

$$\dot{a} = -\partial_a F = -\partial_a e_p^T \Sigma_p^{-1} e_p \tag{2}$$

$$e_p = s_p - g_p(\mu) \tag{3}$$

$$e_v = s_v - g_v(\mu) \tag{4}$$

$$e_f = -f(\mu) \tag{5}$$

Note that this model is a specific instance of the full active inference model [5] tailored to the RHI experiment. We wrote the variational free-energy bound

---

[1] Code will be publicly available at https://github.com/thomasroodnl/active-inference-rhi.

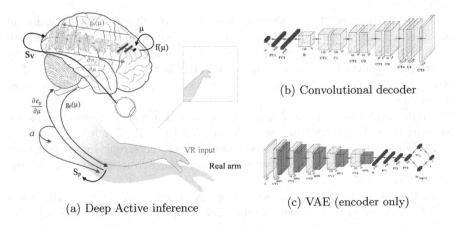

(a) Deep Active inference

(b) Convolutional decoder

(c) VAE (encoder only)

Fig. 1. Deep active inference model for the virtual rubber-hand illusion. (a) The brain variables μ that represent the body state are inferred through proprioceptive e_p and visual e_v prediction errors and their own dynamics $f(\mu)$. During the VR immersion, the agent only sees the VR arm. The ensuing action is driven by proprioceptive prediction errors. The generative visual process is approximated by means of a deep neural network that encodes the sensory input into the body state through a bottleneck. (b, c) Visual generative architectures tested.

in terms of the prediction error e and for clarity, we split it into three terms that correspond to the visual, proprioceptive and dynamical component of the body state. The variances $\Sigma_v, \Sigma_p, \Sigma_\mu$ encode the reliability of the visual, proprioceptive and dynamics information, respectively, that is used to infer the body state. The dynamics of the prediction errors are governed by different generative processes. Here, $g_v(\mu)$ is the generative process of the visual information (i.e. the predictor of the visual input given the brain state variables), $g_p(\mu)$ is the proprioceptive generative process and $f(\mu)$ denotes internal state dynamics (i.e. how the brain variables evolve in time)[2].

Due to the static characteristics of the passive RHI experiment we can simplify the model. First, the generative dynamics model does not affect body update because the experimental setup does not allow for body movement. Second, we fully describe the body state by the joint angles. This means that the s_p and the body state match. Thus, $g(\mu) = \mu$ plus noise and the inverse mapping $\partial_\mu g_p(\mu)$ becomes an all-ones vector. Relaxing these two assumptions is out of the scope of this paper. We can finally write the differential equations with the generative models as follows:

$$\dot{\mu} = \Sigma_p^{-1}(s_p - g_p(\mu)) + \partial_\mu g_v(\mu)^T \gamma \Sigma_v^{-1}(s_v - g_v(\mu)) \tag{6}$$

$$\dot{a} = -\Delta_t \Sigma_p^{-1}(s_p - g_p(\mu)) \tag{7}$$

---

[2] Note that in Eq. (5), the prediction error with respect to the internal dynamics $e_f = \mu' - f(\mu)$ was simplified to $e_f = -f(\mu)$ under the assumption that $\mu' = 0$. In other words, we assume no dynamics on the internal variables.

where γ has been included in the visual term to modulate the level of causality regarding whether the visual information has been produced by our body in the RHI – see Subsect. 2.2. Equation 7 is only valid if the action is the velocity of the joint. Thus, the sensor change given the action corresponds to the time interval between each iteration $\partial_a s = \Delta_t$.

We scale up the model to high-dimensional inputs such as images by approximating the visual generative model $g_v(\mu)$ and the partial derivative of the error with respect to the brain variables $\partial_\mu e_v$ by means of deep neural networks, inspired by [19].

2.1 Generative Model Learning

We learn the forward and inverse generative process of the sensory input by exploiting the representational capacity of deep neural networks. Although in this work we only address the visual input, this method can be extended to any other modality. To learn the visual forward model $g_v(\mu)$ we compare two different deep learning architectures, that is, a convolutional decoder (Fig. 1b) and a variational autoencoder (VAE, Fig. 1c).

The convolutional decoder was designed in similar fashion to the architecture used in [19]. After training the relation between the visual input and the body state, the visual prediction can be computed through the forward pass of the network and its inverse $\partial g(\mu)/\partial \mu$ by means of the backward pass. The VAE was designed using the same decoding structure as the convolutional decoder to allow a fair performance comparison. This means that these models mainly differed in the way they were trained. In the VAE approach we train using the full architecture and we just use the decoder to compute the predictions in the model.

2.2 Modelling Visuo-Tactile Stimulation Synchrony

To synthetically replicate the RHI we need to model both synchronous and asynchronous visuo-tactile stimulation conditions. We define the timepoints at which a visual stimulation event and the corresponding tactile stimulation take place, denoted t_v and t_t respectively. Inspired by the Bayesian causal model [18], we distinguish between two causal explanations of the observed data. That is, $C = c_1$ signifies that the observed (virtual) hand produced both the visual and the tactile events whereas $C = c_2$ signifies that the observed hand produced the visual event and our real hand produced the tactile event (visual and tactile input come from two different sources). The causal impact of the visual information on the body state is represented by

$$\gamma = p(c_1 \mid t_v, t_t) = \frac{p(t_v, t_t \mid c_1)p(c_1)}{p(t_v, t_t \mid c_1)p(c_1) + p(t_v, t_t \mid c_2)p(c_2)} \tag{8}$$

where $p(t_v, t_t \mid c_1)$ is defined as a zero-mean Gaussian distribution over the difference between the timepoints ($p(t_v - t_t \mid c_1)$) and $p(t_v, t_t \mid c_2)$ is defined as a uniform distribution since under c_2, no relation between t_v and t_t is assumed. This yields the update rule

$$\gamma_{t+1} = \begin{cases} \frac{p(t_v,t_t|c_1)\gamma_t}{p(t_v,t_t|\gamma_t)\cdot\gamma_t + p(t_v,t_t|c_2)(1-\gamma_t)} & \text{if visuo-tactile event} \\ \gamma_t \cdot exp(-\frac{(t-max(t_v,t_t))^2}{\Delta_t^{-1}} \cdot r_{decay}), & \text{otherwise} \end{cases} \quad (9)$$

Note that γ is updated only in case of visuo-tactile events. Otherwise, an exponential decay is applied.

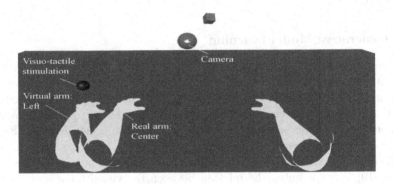

Fig. 2. Virtual environment and experimental setup modelled in the Unity engine.

3 Experimental Setup

We modelled the RHI in a virtual environment created in Unity, as depicted in Fig. 2. This environment was build to closely match the experimental setup used in the human study described in [16]. This experiment exposed human participants to a virtual arm located to the left and right of their real arm, and applied visuo-tactile stimulation by showing a virtual ball touching the hand and applying a corresponding vibration to the hand. Here, the agent's control consisted of two degrees of freedom: shoulder adduction/abduction and elbow flexion/extension. The environment provided proprioceptive information on the shoulder and elbow joint angles to the agent. Visual sensory input to the model originated from a camera located between the left and the right eye position, producing 256×256 pixel grayscale images. Finally, the ML-Agents toolkit was used to interface between the Unity environment and the agent in Python [9]. The agent arm was placed in a forward resting position such that the hand was located 30 cm to the left of the body midline (center position). Three virtual arm location conditions were evaluated: Left, Center and Right. The Center condition matched the information given by proprioceptive input. Visuo-tactile stimulation was applied by generating a visual event at a regular interval of two seconds, followed by a tactile event after a random delay sampled in the range $[0, 0.1)$ for synchronous stimulation and in the range $[0, 1)$ for asynchronous stimulation. The initial γ value was set to 0.01 and we ran $N = 5$ trials each for 30 s (1500 iterations).

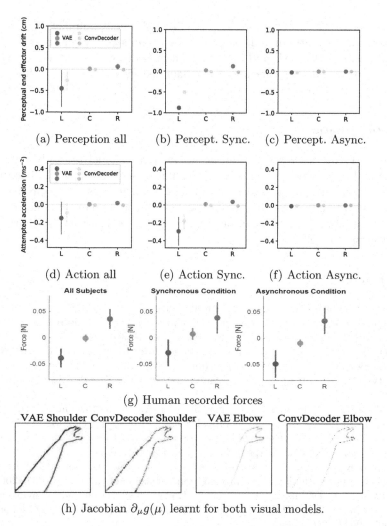

(a) Perception all   (b) Percept. Sync.   (c) Percept. Async.

(d) Action all   (e) Action Sync.   (f) Action Async.

(g) Human recorded forces

(h) Jacobian $\partial_\mu g(\mu)$ learnt for both visual models.

Fig. 3. Model results. (a, b, c) Mean perceptual end-effector drift (in cm). (d,e,f) Mean horizontal end-effector acceleration. (g) Mean forces exerted by human participants in a virtual rubber-hand experiment (from [16]). (h) Visual representation of the Jacobian learnt for the visual models.

4   Results

We observed similar patterns in the drift of the perceived end-effector location (Fig. 3a) and the end-effector action (Fig. 3). These agree with the behavioural data obtained in human experiments (Fig. 3g). For the left and right condition, we observed forces in the direction of the virtual hand during synchronous stimulation (Fig. 3e). However, non-meaningful forces were produced using the convolutional decoder for the right condition. For the center condition, both models produced near-zero average forces. Lastly, asynchronous stimulation produced, with both models, attenuated forces (Fig. 3f). The learnt visual representation differed between the VAE and the Convolutional decoder approaches (Fig. 3h). The VAE obtained smoother and more bounded visual Jacobian values, likely due to its probabilistic latent space.

5   Conclusion

In this work, we described a deep active inference model to study body perception and action during sensorimotor conflicts, such as the RHI. The model, operating as an artificial agent in a virtual environment, was able to produce similar perceptual and active patterns to those found in humans. Further research will address how this model can be employed to investigate the construction of the sensorimotor self [15].

References

1. Asai, T.: Illusory body-ownership entails automatic compensative movement: for the unified representation between body and action. Exp. Brain Res. **233**(3), 777–785 (2014). https://doi.org/10.1007/s00221-014-4153-0
2. Botvinick, M., Cohen, J.: Rubber hands 'feel' touch that eyes see. Nature **391**(6669), 756–756 (1998). https://doi.org/10.1038/35784
3. Botvinick, M., Toussaint, M.: Planning as inference. Trends Cogn. Sci. **16**(10), 485–488 (2012)
4. Buckley, C.L., Kim, C.S., McGregor, S., Seth, A.K.: The free energy principle for action and perception: a mathematical review. J. Math. Psychol. **81**, 55–79 (2017)
5. Friston, K.: The free-energy principle: a unified brain theory? Nat. Rev. Neurosci. **11**(2), 127–138 (2010). https://doi.org/10.1038/nrn2787
6. Friston, K., Mattout, J., Trujillo-Barreto, N., Ashburner, J., Penny, W.: Variational free energy and the laplace approximation. Neuroimage **34**(1), 220–234 (2007)
7. Friston, K.J., Daunizeau, J., Kilner, J., Kiebel, S.J.: Action and behavior: a free-energy formulation. Biol. Cybern. **102**(3), 227–260 (2010). https://doi.org/10.1007/s00422-010-0364-z
8. Hinz, N.A., Lanillos, P., Mueller, H., Cheng, G.: Drifting perceptual patterns suggest prediction errors fusion rather than hypothesis selection: replicating the rubber-hand illusion on a robot. arXiv preprint arXiv:1806.06809 (2018)
9. Juliani, A., et al.: Unity: a general platform for intelligent agents (2018)
10. Kalckert, A., Ehrsson, H.H.: The onset time of the ownership sensation in the moving rubber hand illusion. Front. Psychol. **8**, 344 (2017). https://doi.org/10.3389/fpsyg.2017.00344
11. Kappen, H.J., Gómez, V., Opper, M.: Optimal control as a graphical model inference problem. Mach. Learn. **87**(2), 159–182 (2012). https://doi.org/10.1007/s10994-012-5278-7
12. Kilteni, K., Maselli, A., Kording, K.P., Slater, M.: Over my fake body: body ownership illusions for studying the multisensory basis of own-body perception. Front. Hum. Neurosci. **9**, 141 (2015)
13. Körding, K.P., Wolpert, D.M.: Bayesian integration in sensorimotor learning. Nature **427**(6971), 244–247 (2004). https://doi.org/10.1038/nature02169

14. Lanillos, P., Cheng, G.: Adaptive robot body learning and estimation through predictive coding. In: Proceedings of the IEEE/RSJ International Conference on Intelligent Robots and Systems (IROS 2018), pp. 4083–4090. IEEE (2018)

15. Lanillos, P., Dean-Leon, E., Cheng, G.: Enactive self: a study of engineering perspectives to obtain the sensorimotor self through enaction. In: Joint IEEE International Conference on Developmental Learning and Epigenetic Robotics (2017)

16. Lanillos, P., Franklin, S., Franklin, D.W.: The predictive brain in action: involuntary actions reduce body prediction errors. bioRxiv (2020). https://doi.org/10.1101/2020.07.08.191304

17. Oliver, G., Lanillos, P., Cheng, G.: Active inference body perception and action for humanoid robots. arXiv preprint arXiv:1906.03022 (2019)

18. Samad, M., Chung, A.J., Shams, L.: Perception of body ownership is driven by Bayesian sensory inference. PLoS ONE **10**(2), e0117178–e0117178 (2015). https://doi.org/10.1371/journal.pone.0117178, https://pubmed.ncbi.nlm.nih.gov/25658822

19. Sancaktar, C., van Gerven, M., Lanillos, P.: End-to-end pixel-based deep active inference for body perception and action. arXiv preprint arXiv:2001.05847 (2020)

You Only Look as Much as You Have To
Using the Free Energy Principle for Active Vision

Toon Van de Maele[✉], Tim Verbelen, Ozan Çatal, Cedric De Boom,
and Bart Dhoedt

IDLab, Department of Information Technology,
Ghent University – imec, Ghent, Belgium
toon.vandemaele@ugent.be

Abstract. Active vision considers the problem of choosing the optimal
next viewpoint from which an autonomous agent can observe its environment. In this paper, we propose to use the active inference paradigm as a
natural solution to this problem, and evaluate this on a realistic scenario
with a robot manipulator. We tackle this problem using a generative
model that was learned unsupervised purely from pixel-based observations. We show that our agent exhibits information-seeking behavior,
choosing viewpoints of regions it has not yet observed. We also show
that goal-seeking behavior emerges when the agent has to reach a target
goal, and it does so more efficiently than a systematic grid search.

Keywords: Active vision · Active inference · Deep generative
modelling · Robotic planning

1 Introduction

Active vision considers an observer that can act by controlling the geometric
properties of the sensor in order to improve the quality of the perceptual results
[1]. This problem becomes apparent when considering occlusions, a limited field
of view or a limited resolution of the used sensor [2]. In many cases, selecting the next viewpoint should be done as efficiently as possible due to limited
resources for processing the new observations and the time it takes to reach
the new observation pose. This problem is traditionally solved with frontier-
based methods [21] in which the environment is represented as an occupancy
grid. These approaches rely on evaluating engineered utility functions that estimate the amount of new information provided for all potential viewpoints [8,21].
Usually this utility function represents the amount of unobserved voxels that a
given viewpoint will uncover. Instead of using hand-crafted heuristics, this function can also be learned from data [8,9]. A different approach is to predict the
optimal viewpoint with respect to reducing uncertainty and ambiguity directly
from a reconstructed volumetric grid [3,13]. A different bio-inspired method for
active vision is proposed by Rasouli et al. [17] in which the action is driven by a

© Springer Nature Switzerland AG 2020
T. Verbelen et al. (Eds.): IWAI 2020, CCIS 1326, pp. 92–100, 2020.
https://doi.org/10.1007/978-3-030-64919-7_11

visual attention mechanism in conjunction with a non-myopic decision-making algorithm that takes previous observations at different locations in account.

Friston et al. [7,14] cast the active vision problem as a low dimensional, discrete state-space Markov decision process (MDP) that can be solved using the active inference framework. In this paradigm, agents act in order to minimize their surprise, i.e. their free energy. In this paper, instead of using an explicit 3D representation, or a simple MDP formulation of the environment, we learn a generative model and latent state distribution purely from observations. Previous work also used deep learning techniques to learn the generative model in order to engage in active inference [22], while other work has created an end-to-end active inference pipeline using pixel-based observations [20]. Similar to Friston et al. [6,7,14], we then use the expected free energy to drive action selection. Similar to the work of Nair, Pong et al. [15] where the imagined latent state is used to compute the reward value for optimizing reinforcement learning tasks and the work of Finn and Levine [5], where a predictive model is used that estimates the pixel observations for different control policies, we employ the imagined observations form the generative model to compute the expected free energy. We evaluate our method on a grasping task with a robotic manipulator with an in-hand camera. In this task, we want the robot to get to the target object as fast as possible. For this reason we consider the case of best viewpoint selection. We show how active inference yields information-seeking behavior, and how the robot is able to reach goals faster than random or systematic grid search.

2   Active Inference

Active inference posits that all living organisms minimize free energy (FE) [6]. The variational free energy is given by:

$$
\begin{aligned}
F &= \mathbb{E}_Q[\log Q(\tilde{\mathbf{s}}) - \log P(\tilde{\mathbf{o}}, \tilde{\mathbf{s}}, \pi)] \\
&= D_{KL}[Q(\tilde{\mathbf{s}}) \| P(\tilde{\mathbf{s}}, \pi)] - \mathbb{E}_Q[\log P(\tilde{\mathbf{o}} | \tilde{\mathbf{s}}, \pi)],
\end{aligned}
\tag{1}
$$

where $\tilde{\mathbf{o}}$ is a sequence of observations, $\tilde{\mathbf{s}}$ the sequence of corresponding model belief states, π the followed policy or sequence of actions taken, and $Q(\tilde{\mathbf{s}})$ the approximate posterior of the joint distribution $P(\tilde{\mathbf{o}}, \tilde{\mathbf{s}}, \pi)$. Crucially, in active inference, policies are selected that minimize the expected free energy $G(\pi, \tau)$ for future timesteps τ [6]:

$$
G(\pi, \tau) \approx -\mathbb{E}_{Q(\mathbf{o}_\tau | \pi)}[D_{KL}[Q(\mathbf{s}_\tau | \mathbf{o}_\tau, \pi) \| Q(\mathbf{s}_\tau | \pi)]] - \mathbb{E}_{Q(\mathbf{o}_\tau | \pi)}[\log P(\mathbf{o}_\tau)]. \tag{2}
$$

This can be viewed as a trade-off between an epistemic, uncertainty-reducing term and an instrumental, goal-seeking term. The epistemic term is the Kullback-Leibler divergence between the expected future belief over states when following policy π and observing $\mathbf{o}_\tau$ and the current belief. The goal-seeking term is the likelihood that the goal will be observed when following policy π.

3 Environment and Approach

In this paper, we consider a simulated robot manipulator with an in-hand camera which can actively query observations from different viewpoints or poses by moving its gripper, as shown in Fig. 1. The robotic agent acts in a static workspace, in which we randomly spawn a red, green and blue cube of fixed size. Each such configuration of random cube positions is dubbed a scene, and the goal of the robot is to find a cube of a particular color in the workspace. The agent initially has no knowledge about the object positions and has to infer this information from multiple observations at different poses. Example observations for different downward facing poses are given in Fig. 2.

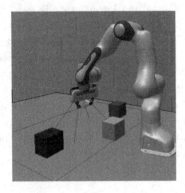
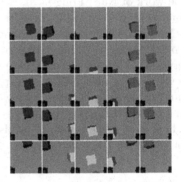

Fig. 1. Franka Emika Panda robot in the CoppeliaSim simulator in a random scene with three colored cubes. (Color figure online)

Fig. 2. Sampled observations on a grid of potential poses used for evaluating the expected free energy. (Color figure online)

To engage in active inference, the agent needs to be equipped with a generative model. This generative model should be able to generate new observations given an action or in this particular case, the new robot pose. In contrast with [7,14], we do not fix the generative model upfront, but learn it from data. We generate a dataset of 250 different scenes consisting of approximately 25 discrete time steps in which the robot observes the scene from a different viewpoint. Using this dataset we train two deep neural networks to approximate the likelihood distribution $P(\mathbf{o}_t|\mathbf{s}_t, \pi)$ and approximate posterior distribution $Q(\mathbf{s}_t|\mathbf{o}_t, \pi)$ as multivariate Gaussian distributions. In our notation, $\mathbf{o}_t$ and $\mathbf{s}_t$ respectively represent the observation and latent state at discrete timestep t. Both distributions are conditioned by the policy π, the action that the robot should take in order to acquire a new observation, or equivalently, the new observation viewpoint. The models are optimized by minimizing the free energy from Eq. 1, with a zero-mean isotropic Gaussian prior $P(\mathbf{s}_t|\pi) = \mathcal{N}(0, 1)$. Hence the system is trained as an encoder-decoder to predict scene observations from unseen poses, given a number of observations from the same scene at different poses. This is similar to a Generative Query Network (GQN) [4]. For more details on the model architecture and training hyperparameters, we refer to Appendix A.

At inference time, the policy π, or equivalently the next observer pose, is selected by evaluating Eq. (2) for a number of candidate policies and selecting the policy that evaluates to the lowest expected free energy. These candidate policies are selected by sampling a grid of poses over the workspace. The trained decoder extracts the imagined observation for each of the candidate policies and the state vector acquired through encoding the initial observations. The corresponding expected posterior distributions are computed by forwarding these imagined observations together with the initial observations through the encoder. For the goal-seeking term, we provide the robot with a preferred observation, i.e. the image of the colored cube to fetch, and we evaluate $\log P(o_\tau)$. The epistemic term is evaluated by using the likelihood model to imagine what the robot would see from the candidate pose, and then calculating the KL divergence between the state distributions of the posterior model before and after "seeing" this imagined observation. The expectation terms are approximated by drawing a number of samples for each candidate pose.

4   Experiments

We evaluate our system in two scenarios. In the first scenario, only the epistemic term is taken into account, which results in an exploring agent that actively queries information of the scene. In the second scenario, we add the instrumental term by which the agent makes an exploration-exploitation trade-off to reach the goal state as fast as possible.

4.1   Exploring Behaviour

First, we focus on exploratory or information-seeking behaviour, i.e. actions are chosen based on the minimization of only the epistemic term of the expected free energy. For evaluation we restrict the robot arm to a fixed number of poses at a fixed height close to the table, so it can only observe a limited area of the workspace. The ground truth observations corresponding to the candidate poses are shown in a grid in Fig. 2.

Initially, the agent has no information about the scene, and the initial state distribution $Q(s)$ is a zero-mean isotropic Gaussian. The expected observation is computed over 125 samples and visualized in the top row of Fig. 3a. Clearly, the agent does not know the position of any of the objects in the scene, resulting in a relatively low value of the epistemic term from Eq. (2) for all candidate poses. This is plotted in the bottom row of Fig. 3a. The agent selects the upper left pose as indicated by the green hatched square in Fig. 3b. After observing the blue cube in the upper left corner, the epistemic value of the left poses drops, and the robot queries a pose at the right side of the workspace. Finally, the robot queries one of the central poses, and the epistemic value of all poses becomes relatively high, as new observations do not yield more information. Notice that at this point, the robot can also accurately reconstruct the correct cubes from any pose as shown in the top row of Fig. 3d.

4.2 Goal Seeking Behaviour

In this experiment, we use the same scene and grid of candidate poses, but now we provide the robot with a preferred observation from the red cube, indicated by the red hatched square in the bottom row of Figs. 4a through 4d.

Initially, the agent has no information on the targets position and the same information-seeking behaviour from Sect. 4.1 can be observed in the first steps as the epistemic value takes the upper hand. However, after the second step, the agent has observed the red cube and knows which pose will reach the preferred state. The instrumental value takes the upper hand as indicated by the red values in Figs. 4a through 4d. This is reflected by a significantly lower expected free energy. Even though the agent has not yet observed the green cube and is unable to create correct reconstructions as shown in Fig. 4d, it will drive itself towards the preferred state. The trade off between exploratory and goal seeking behaviour can clearly be observed. In Fig. 4c, the agent still has low epistemic values for the candidate poses to the left, but they do not outweigh the low instrumental value to reach the preferred state. The middle column of potential observations has a lower instrumental value, which is the result of using independent Gaussians for estimating likelihood on each pixel.

The number of steps to reach the preferred state is computed on 30 different validation scenes not seen in training, where the preferred state is chosen randomly. On average, the agent needs 3.7 steps to reach its goal. This is clearly more efficient than a systematic grid search which would take on average 12.5 steps.

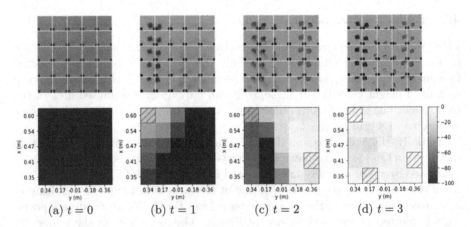

(a) $t = 0$ (b) $t = 1$ (c) $t = 2$ (d) $t = 3$

Fig. 3. The top row represents the imagined observations, i.e. the observations generated by the generative model, for each of the considered potential poses at a given step, the bottom row represents the epistemic value for the corresponding poses. Darker values represent a larger influence of the epistemic value. The green hatched squares mark the observed poses. (Color figure online)

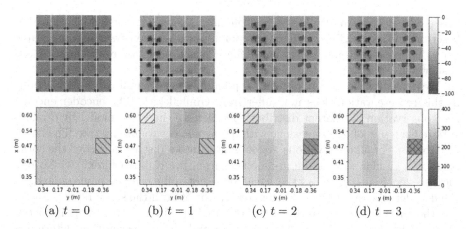

(a) $t = 0$ (b) $t = 1$ (c) $t = 2$ (d) $t = 3$

Fig. 4. The top row shows the imagined observations for each of the considered potential poses at a given time step. The bottom row shows the expected free energy for the corresponding poses. Blue is used to represent the epistemic value, while red is used to represent the instrumental value. The values of both terms are shown in the legend. The green hatched squares mark the observed poses, while the red hatched square marks the preferred state. (Color figure online)

5 Conclusion

This work shows promising results in using the active inference framework for active vision. The problem is tackled with a generative model learned unsupervised from pure pixel data. The proposed approach can be used for efficiently exploring and solving robotic grasping scenarios in complex environments where a lot of uncertainty is present, for example in cases with a limited field of view or with many occlusions.

We show that it is possible to use learned latent space models as generative models for active inference. We show that both exploring and goal-seeking behaviour surfaces when using active inference as an action-selection policy. We demonstrated our approach in a realistic robot simulator and plan to extend this to a real world setup as well.

Acknowledgments. This research received funding from the Flemish Government under the "Onderzoeksprogramma Artificiële Intelligentie (AI) Vlaanderen" programme.

A   The Generative Model

The generative model, described in this paper, is approximated by a neural network that predicts a multivariate Gaussian distribution with a diagonal covariance matrix. We consider a neural network architecture from the family of the variational autoencoders (VAE) [11,18] which is very similar to the Generative Query Network (GQN) [4]. In contrast to the traditional autoencoders, this model encodes multiple observations into a single latent distribution that describes the scene. Given a query viewpoint, new unseen views can be generated from the encoded scene description. A high level description of the architecture is shown in Fig. 5.

We represent the camera pose as 3D point and the orientation as a quaternion as this representation does not suffer from Gimbal lock. The encoder encodes each observation in a latent distribution which we choose to model by a multivariate Gaussian of 32 dimensions with a diagonal covariance matrix. The latent distributions of all observations are combined into a distribution over the entire scene in a similar way as the update step from the Kalman filter [10]. No prediction step is necessary as the agent does not influence the environment. In the decoder, the input is a concatenated vector of both the scene representation and the query viewpoint. Intuitively, both are important as the viewpoint determines which area of the scene is observed and the representation determines which objects are visible at each position. Between the convolutional layers, the intermediate representation is transformed using a FiLM layer, conditioned on the input vector, this allows the model to learn which features are relevant at different stages of the decoding process.

A dataset of 250 scenes, each consisting of approximately 25 (image, viewpoint) pairs has been created in a simulator in order to train this model. To limit the complexity of this model, all observations consist of the same fixed downward orientation.

Table 1. Training implementation details.

Optimizer	Adam
Learning rate	0.0001
Batch size	10
Number of observations	3–10
Tolerance	75.0
λ_{max}	100.0
λ_{init}	20.0

The neural network is optimized using the Adam optimizer algorithm with parameters shown in Table 1. For each scene between 3 and 10 randomly picked observations are provided to the model, from which it is tasked to predict a new one. The model is trained end-to-end using the GECO algorithm [19] on the following loss function:

$$\mathcal{L}_\lambda = D_{KL}[Q(\tilde{s}|\bar{o})||\mathcal{N}(0,I)] + \lambda \cdot \mathcal{C}(o,\hat{o}) \tag{3}$$

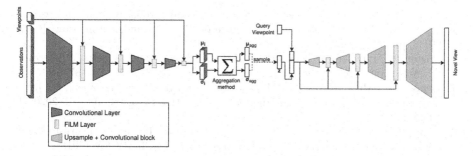

Fig. 5. Schematic view of the generative model. The left part is the encoder that produces a latent distribution for every observation, viewpoint pair. This encoder consists of 4 convolutional layers interleaved with FiLM [16] layers that condition on the viewpoint. This transforms the intermediate representation to encompass the spatial information from the viewpoint. The latent distributions are combined to form an aggregated distribution over the latent space. A sampled vector is concatenated with the query viewpoint from which the decoder generates a novel view. The decoder mimicks the encoder architecture and has 4 convolutional cubes (upsamples the image and processes it with two convolutional layers) interleaved with a FiLM layer that conditions on the concatenated information vector. Each layer is activated with a LeakyReLU [12] activation function.

The constraint $\mathcal{C}$ is applied to a MSE loss on the reconstructed and ground truth observation. This constraint simply means that the MSE should stay below a fixed tolerance. λ is a Lagrange multiplier and the loss is optimized using a min-max scheme [19]. Specific implementation values are shown in Table 1.

The expected free energy is computed for a set of potential poses. The generative model is first used to estimate the expected view for each considered pose. The expected value of the posterior with this expected view is computed for a large number of samples. This way, the expected epistemic term is computed. For numerical stability, we clamp the variances of the posterior distributions to a value of 0.25. The instrumental value is computed as the MSE between the preferred state and the expected observation. This essentially boils down to computing the log likelihood of every pixel is modelled by a Gaussian with a fixed variance of 1.

References

1. Aloimonos, J., Weiss, I., Bandyopadhyay, A.: Active vision. Int. J. Comput. Vision **1**(4), 333–356 (1988). https://doi.org/10.1007/bf00133571
2. Denzler, J., Zobel, M., Niemann, H.: Information theoretic focal length selection for real-time active 3D object tracking. In: ICCV (2003)
3. Doumanoglou, A., Kouskouridas, R., Malassiotis, S., Kim, T.K.: 6D object detection and next-best-view prediction in the crowd. In: CVPR (2016)
4. Ali Eslami, S.M., et al.: Neural scene representation and rendering. Science **360**, 1204–1210 (2018)

5. Finn, C., Levine, S.: Deep visual foresight for planning robot motion. In: Proceedings of the IEEE International Conference on Robotics and Automation (ICRA 2017), pp. 2786–2793 (2017)
6. Friston, K., FitzGerald, T., Rigoli, F., Schwartenbeck, P., O'Doherty, J., Pezzulo, G.: Active inference and learning. Neurosci. Biobehav. Rev. **68**, 862–879 (2016)
7. Heins, R.C., Mirza, M.B., Parr, T., Friston, K., Kagan, I., Pooresmaeili, A.: Deep active inference and scene construction (2020). https://doi.org/10.1101/2020.04.14.041129
8. Hepp, B., Dey, D., Sinha, S.N., Kapoor, A., Joshi, N., Hilliges, O.: Learn-to-score: efficient 3D scene exploration by predicting view utility. In: ECCV (2018)
9. Kaba, M.D., Uzunbas, M.G., Lim, S.N.: A reinforcement learning approach to the view planning problem. In: CVPR (2017)
10. Kalman, R.E.: A new approach to linear filtering and prediction problems. Trans. ASME J. Fluids Eng. **82**(1), 35–45 (1960). https://doi.org/10.1115/1.3662552
11. Kingma, D.P., Welling, M.: Auto-encoding variational Bayes. In: Proceedings of the 2nd International Conference on Learning Representations, ICLR 2014 - Conference Track Proceedings (Ml), pp. 1–14 (2014)
12. Maas, A.L., Hannun, A.Y., Ng, A.Y.: Rectifier nonlinearities improve neural network acoustic models. In: ICML Workshop on Deep Learning for Audio, Speech and Language Processing (2013)
13. Mendoza, M., Vasquez-Gomez, J.I., Taud, H., Sucar, L.E., Reta, C.: Supervised learning of the next-best-view for 3D object reconstruction. Pattern Recogn. Lett. **133**, 224–231 (2020)
14. Mirza, M.B., Adams, R.A., Mathys, C.D., Friston, K.J.: Scene construction, visual foraging, and active inference. Front. Comput. Neurosci. **10**, 56 (2016)
15. Nair, A.V., Pong, V., Dalal, M., Bahl, S., Lin, S., Levine, S.: Visual reinforcement learning with imagined goals. In: Bengio, S., Wallach, H., Larochelle, H., Grauman, K., Cesa-Bianchi, N., Garnett, R. (eds.) Advances in Neural Information Processing Systems, vol. 31, pp. 9191–9200. Curran Associates, Inc. (2018). http://papers.nips.cc/paper/8132-visual-reinforcement-learning-with-imagined-goals.pdf
16. Perez, E., Strub, F., De Vries, H., Dumoulin, V., Courville, A.: FiLM: visual reasoning with a general conditioning layer. In: Proceedings of the 32nd AAAI Conference on Artificial Intelligence, AAAI 2018, pp. 3942–3951 (2018)
17. Rasouli, A., Lanillos, P., Cheng, G., Tsotsos, J.K.: Attention-based active visual search for mobile robots. Auton. Robots **44**(2), 131–146 (2019). https://doi.org/10.1007/s10514-019-09882-z
18. Rezende, D.J., Mohamed, S., Wierstra, D.: Stochastic backpropagation and approximate inference in deep generative models. In: Proceedings of the 31st International Conference on Machine Learning, ICML 2014, vol. 4, pp. 3057–3070 (2014)
19. Rezende, D.J., Viola, F.: Taming vaes. CoRR abs/1810.00597 (2018)
20. Sancaktar, C., van Gerven, M., Lanillos, P.: End-to-end pixel-based deep active inference for body perception and action (2019)
21. Yamauchi, B.: A frontier-based exploration for autonomous exploration. In: ICRA (1997)
22. Çatal, O., Verbelen, T., Nauta, J., Boom, C.D., Dhoedt, B.: Learning perception and planning with deep active inference. In: ICASSP, pp. 3952–3956 (2020)

Modulation of Viability Signals for Self-regulatory Control

Alvaro Ovalle[✉] and Simon M. Lucas

Queen Mary University of London, London, UK
{a.ovalle,simon.lucas}@qmul.ac.uk

Abstract. We revisit the role of instrumental value as a driver of adaptive behavior. In active inference, instrumental or extrinsic value is quantified by the information-theoretic *surprisal* of a set of observations measuring the extent to which those observations conform to prior beliefs or preferences. That is, an agent is expected to seek the type of evidence that is consistent with its own model of the world. For reinforcement learning tasks, the distribution of preferences replaces the notion of reward. We explore a scenario in which the agent learns this distribution in a self-supervised manner. In particular, we highlight the distinction between observations induced by the environment and those pertaining more directly to the continuity of an agent in time. We evaluate our methodology in a dynamic environment with discrete time and actions. First with a surprisal minimizing model-free agent (in the RL sense) and then expanding to the model-based case to minimize the expected free energy.

Keywords: Perception-action loop · Active inference · Reinforcement learning · Self-regulation · Anticipatory systems · Instrumental value

1 Introduction

The continual interaction that exists between an organism and the environment requires an active form of regulation of the mechanisms safeguarding its integrity. There are several aspects an agent must consider, ranging from assessing various sources of information to anticipating changes in its surroundings. In order to decide what to do, an agent must consider between different courses of action and factor in the potential costs and benefits derived from its hypothetical future behavior. This process of selection among different value-based choices can be formally described as an optimization problem. Depending on the formalism, the cost or utility functions optimized by the agent presuppose different normative interpretations.

In reinforcement learning (RL) for instance, an agent has to maximize the expected reward guided by a signal provided externally by the environment in an oracular fashion. The reward in some cases is also complemented with an *intrinsic* contribution, generally corresponding to an epistemic deficiency within

© Springer Nature Switzerland AG 2020
T. Verbelen et al. (Eds.): IWAI 2020, CCIS 1326, pp. 101–113, 2020.
https://doi.org/10.1007/978-3-030-64919-7_12

the agent. For example prediction error [24], novelty [3,5,23] or ensemble disagreement [25]. It is important to note that incorporating these surrogate rewards into the objectives of an agent is often regarded as one of many possible enhancements to increase its performance, rather than been motivated by a concern with explaining the roots of goal-directed behavior.

In active inference [14], the optimization is framed in terms of the minimization of the *variational free energy* to try to reduce the difference between sensations and predictions. Instead of rewards, the agent holds a prior over preferred future outcomes, thus an agent minimizing its free energy acts to maximize the occurrence of these preferences and to minimize its own *surprisal*. Value arises not as an external property of the environment, but instead it is conferred by the agent as a contextual consequence of the interplay of its current configuration and the interpretation of stimuli.

There are recent studies that have successfully demonstrated how to reformulate RL and control tasks under the active inference framework. While for living processes it is reasonable to assume that the priors emerge and are refined over evolutionary scales and during a lifetime, translating this view into a detailed algorithmic characterization raises important considerations because there is no evolutionary prior to draw from. Thus the approaches to specify a distribution of preferences have included for instance, taking the reward an RL agent would receive and encoding it as the prior [16,21,29,32–34], connecting it to task objectives [29] or through expert demonstrations [6,7,30].

In principle this would suggest that much of the effort that goes into reward engineering in RL is relocated to that of specifying preferred outcomes or to the definition of a phase space. Nonetheless active inference provides important conceptual adjustments that could potentially facilitate conceiving more principled schemes towards a theory of agents that could provide a richer account of autonomous behavior and self-generation of goals, desires or preferences. These include the formulation of objectives and utilities under a common language residing in belief space, and appealing to a worldview in which utility is not treated as independent or detached from the agent. In particular the latter could encourage a more organismic perspective of the agent in terms of the perturbations it must endure and the behavioral policies it attains to maintain its integrity [11].

Here we explore this direction by considering how a signal acquires functional significance as the agent identifies it as a condition necessary for its viability and future continuity in the environment. Mandated by an imperative to minimize surprisal, the agent learns to associate sensorimotor events to specific outcomes. First, we start by introducing the surprise minimizing RL (SMiRL) specification [4] before we proceed with a brief overview of the expected free energy. Then we motivate our approach from the perspective of a self-regulatory organism. Finally, we present results from our case study and close with some observations and further potential directions.

2   Preliminaries

2.1   Model-Free Surprisal Minimization

Consider an environment whose generative process produces a state $s_t \in \mathcal{S}$ at each time step t resulting in an agent observing $o_t \in \mathcal{O}$. The agent acts on the environment with $a_t \in \mathcal{A}$ according to a policy π, obtaining the next observation o_{t+1}. Suppose the agent performs density estimation on the last $t-k$ observations to obtain a current set of parameter(s) θ_t summarizing $p_\theta(o)$. As these sufficient statistics contain information about the agent-environment coupling, they are concatenated with the observations into an augmented state $x_t = (o_t, \theta_t)$. Every time step, the agent computes the surprisal generated by a new observation given its current estimate and then updates it accordingly. In order to minimize surprisal under this model-free RL setting, the agent should maximize the expected log of the model evidence $\mathbb{E}[\sum_t \gamma^t \ln p_{\theta_t}(o_t)]$ [4]. Alternatively, we maintain consistency with active inference by expressing the optimal surprisal Q-function as,

$$Q_{\pi^*}(x, a) = \mathbb{E}_\pi[-\ln p_\theta(o) + \gamma \min_{a'} Q_{\pi^*}(x', a')] \tag{1}$$

estimated via DQN [22] or any function approximator with parameters ϕ such that $Q_{\pi^*}(x, a) \approx Q(x, a; \phi)$.

2.2   Expected Free Energy

The free energy principle (FEP) [15] has evolved from an account of message passing in the brain to propose a probabilistic interpretation of self-organizing phenomena [13, 27, 28]. Central to current discourse around the FEP is the notion of the Markov blanket to describe a causal separation between the internal states of a system from external states, as well as the interfacing blanket states (i.e. sensory and active states). The FEP advances the view that a system remains far from equilibrium by maintaining a low entropy distribution over the states it occupies during its lifetime. Accordingly, the system attempts to minimize the surprisal of an event at a particular point in time.

This can be more concretely specified if we consider a distribution $p(o)$ encoding the states, drives or desires the system should fulfil. Thus the system strives to obtain an outcome o that minimizes the surprisal $-\ln p(o)$. Alternatively, we can also state this as the agent maximizing its model evidence or marginal likelihood $p(o)$. For most cases estimating the actual marginal is intractable, therefore a system instead minimizes the free energy [10, 18] which provides an upper bound on the log marginal [19],

$$\mathbf{F} = \mathbb{E}_{q(s)}[\ln q(s) - \ln p(o, s)] \tag{2}$$

where $p(o, s)$ is the generative model and $q(s)$ the variational density approximating hidden causes. Equation 2 is used to compute a static form of free energy and infer hidden causes given a set of observations. However if we instead consider an agent that acts over an extended temporal dimension, it must infer and select policies that minimize the expected free energy (EFE) **G** [14] of a policy π for a future step $\tau > t$. This can be expressed as,

$$\mathbf{G}(\pi, \tau) = \mathbb{E}_{q(o_\tau, s_\tau | \pi)}[\ln q(s_\tau | \pi) - \ln p(o_\tau, s_\tau | \pi)] \tag{3}$$

where $p(o_\tau, s_\tau | \pi) = q(s_\tau | o_\tau, \pi)p(o_\tau)$ is the generative model of the future. Rearranging **G** as,

$$\mathbf{G}(\pi, \tau) = - \underbrace{\mathbb{E}_{q(o_\tau | \pi)}[\ln p(o_\tau)]}_{instrumental\ value} - \underbrace{\mathbb{E}_{q(o_\tau | \pi)}\big[D_{KL}[\ln q(s_\tau | o_\tau, \pi) || \ln q(s_\tau | \pi)]\big]}_{epistemic\ value} \tag{4}$$

which illustrates how the EFE entails a pragmatic, instrumental or goal-seeking term that realizes preferences and an epistemic or information seeking term that resolves uncertainty. An agent selects a policy with probability $q(\pi) = \sigma(-\beta \sum_\tau \mathbf{G}_\tau(\pi))$ where σ is the softmax function and β is the inverse temperature. In summary, an agent minimizes its free energy via active inference by changing its beliefs about the world or by sampling the regions of the space that conforms to its beliefs.

3   Adaptive Control via Self-regulation

The concept of homeostasis has played a crucial role in our understanding of physiological regulation. It describes the capacity of a system to maintain its internal variables within certain bounds. Recent developments in the FEP describing the behavior of self-organizing systems under the framework, can be interpreted as an attempt to provide a formalization of this concept [28]. From this point of view, homeostatic control in an organism refers to the actions necessary to minimize the surprisal of the values reported by interoceptive channels, constraining them to those favored by a viable set of states. Something that is less well understood is how these attracting states come into existence. That is, how do they emerge from the particular conditions surrounding the system and how are they discovered among the potential space of signals.

Recently, it has been shown that complex behavior may arise by minimizing surprisal in observation space (i.e. sensory states) without pre-encoded fixed prior distributions in large state spaces [4]. Here we consider an alternative angle intended to remain closer to the homeostatic characterization of a system. In our scenario, we assume that given the particular dynamics of an environment, if an agent is equipped only with a basic density estimation capacity, then structuring its behavior around the type of regularities in observation space that can sustain it in time will be difficult. In these situations with fast changing dynamics, rather than minimizing free energy over sensory signals, the agent may instead leverage them to maintain a low future surprisal of another target variable. That implies that although the agent may have in principle access to multiple signals it might be interested in maintaining only some of them within certain expected range.

Defining what should constitute the artificial physiology in simulated agents is not well established. Therefore we assume the introduction of an information channel representing in abstract terms the interoceptive signals that inform the agent about its continuity in the environment. We can draw a rudimentary comparison, and think of this value in a similar way in which feelings agglutinate and coarse-grain the changes of several internal physical responses [9]. In addition, we are interested in the agent learning to determine whether it is conductive to its self-preservation in the environment or not.

3.1 Case Study

We assess the behavior of an agent in the *Flappy Bird* environment (Fig. 1 left). This is a task where a bird must navigate between obstacles (pipes) at different positions while stabilizing its flight. Despite the apparent simplicity, the environment offers a fundamental aspect present in the physical world. Namely, the inherent dynamics leads spontaneously to the functional disintegration of the agent. If the agent stops propelling, it succumbs to gravity and falls. At the same time the environment has a constant scrolling rate, which implies that the agent cannot remain floating at a single point and cannot survive simply by flying aimlessly. Originally, the task provides a reward every time the bird traverses in between two pipes. However for our case study the information about the rewards is never propagated and therefore does not have any impact on the behavior of the agent. The agent receives a feature vector of observations indicating its location and those of the obstacles. In addition, the agent obtains a *measurement* v indicating its presence in the task (i.e. 1 or 0). This measurement does not represent anything positive or negative by itself, it is simply another signal that we assume the agent is able to calculate. Similarly to the outline in Sect. 2.1, the agent monitors the last $t - k$ values of this measurement and estimates the density to obtain θ_t. These become the statistics describing the current approximated distribution of preferences $p(v|\theta_t)$ or $p_{\theta_t}(v)$, which are also used to augment the observations to $x_t = (o_t, \theta_t)$. When the agent takes a new measurement v_t, it evaluates the surprisal against $p_{\theta_{t-1}}(v_t)$. In this particular case it is evaluated via a Bernoulli density function such that $-\ln p_{\theta_{t-1}}(v_t) = -(v_t \ln \theta_{t-1} + (1 - v_t) \ln(1 - \theta_{t-1}))$. First, we train a baseline model-free surprisal minimizing DQN as specified in Sect. 2.1 parameterized by a neural network (NN). Then we examine the behavior of a second agent that minimizes the expected free energy. Thus the agent learns an augmented state transition model of the world, parameterized by an ensemble of NNs, and an expected surprisal model, also parameterized by another NN. In order to identify an optimal policy we apply rolling horizon evolution [26] to generate candidate policies $\pi = (a_\tau, ..., a_T)$ and to associate them to an expected free energy given by (Appendix A)

$$\mathbf{G}(\pi, \tau) \approx -\mathbb{E}_{q(o_\tau, v_\tau, \theta | \pi)} D_{KL}[q(s_\tau |, o_\tau, v_\tau, \pi) || q(s_\tau | \pi)] - \mathbb{E}_{q(v_\tau, \theta, s_\tau | \pi)}[\ln p_\theta(v_\tau)]$$
(5)

If we explicitly consider the model parameters ϕ, Eq. 5 can be decomposed as (Appendix B),

$$
\mathbf{G}(\pi, \tau) \approx - \underbrace{\mathbb{E}_{q(o_\tau, v_\tau, \phi | \pi)} D_{KL}[q(s_\tau | o_\tau, v_\tau, \pi) \| q(s_\tau | \pi)]}_{salience}
$$
$$
- \underbrace{\mathbb{E}_{q(o_\tau, v_\tau, s_\tau | \pi)} D_{KL}[q(\phi | s_\tau, o_\tau, v_\tau, \pi) \| q(\phi)]}_{novelty}
$$
$$
- \underbrace{\mathbb{E}_{q(o_\tau, v_\tau, s_\tau, \phi | \pi)}[\ln p_\theta(v_\tau)]}_{instrumental\ value}
$$

(6)

The expression unpacks further the epistemic contributions to the EFE in terms of salience and novelty [17]. These terms refer to the expected reduction in uncertainty about hidden causes and in the parameters respectively. For this task $o = s$, thus only the first and third term are considered.

3.2   Evaluation

The plot on Fig. 1 (center) tracks the performance of an EFE agent in the environment (averaged over 10 seeds). The dotted line represents the surprisal minimizing DQN agent after 1000 episodes. The left axis corresponds to the (unobserved) task reward while the right axis indicates the approximated number of time steps the agent survives. During the first trials, and before the agent exhibits any form of competence, it was observed that the natural coupling between agent and environment grants the agent a life expectancy of roughly 19–62 time steps in the task. This is essential as it starts to populate the statistics of v. Measuring a specific quantity v, although initially representing just another signal, begins to acquire certain value due to the frequency that it occurs. In turn, this starts to dictate the preferences of the agent as it hints that measuring certain signal correlates with having a stable configuration for this particular environment as implied by its low surprisal. Right Fig. 1 shows the evolution of parameter θ (averaged within an episode) corresponding to the distribution of preferred measurements $p_\theta(v)$ which determines the level of surprisal assigned when receiving the next v. As the agent reduces its uncertainty about the environment it also becomes more capable of associating sensorimotor events to specific measurements. The behavior becomes more consistent with seeking less surprising measurements, and as we observe, this reinforces its preferences, exhibiting the circular self-evidencing dynamics that characterize an agent minimizing its free energy.

4   Discussion

Learning Preferences in Active Inference: The major thesis in active inference is the notion of an agent acting in order to minimize its expected surprise.

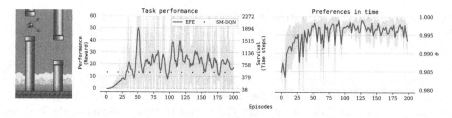

Fig. 1. *Left*: The Flappy Bird environment. *Center*: Performance of an EFE agent. The left axis indicates the unobserved rewards as reported by the task and the right axis the number of time steps it survives in the environment. The dotted line shows the average performance of an SM-DQN after 1000 episodes. *Right*: Parameter θ in time, summarizing the intra-episode sufficient statistics of $p_\theta(v)$.

This implies that the agent will exhibit a tendency to seek for the sort of outcomes that have high prior probability according to a biased model of the world, giving rise to goal-directed behavior. Due to the difficulty of modeling an agent to exhibit increasing levels of autonomy, the agent based simulations under this framework, and similarly to how it has largely occurred in RL, have tended to concentrate on the generation of a particular expected behavior in the agent. That is, on how to make the agent perform a task by encoding predefined goals [16,21,29,32–34] or providing guidance [6,7,30]. However there has been recent progress trying to mitigate this issue. For example, in some of the simulations in [29] the authors included a distribution over prior preferences to account for each of the cells in *Frozen Lake*, a gridworld like environment. Over time the prior preferences are tuned, leading to habit formation. Most related to our work, are the studies on surprise minimizing RL (SMiRL) by [4], where model-free agents performed density estimation on their observation space and acquired complex behavior in various tasks by maximizing the model evidence of their observations. Here we have also opted for this approach, however we have grounded it on organismic based considerations of viability as inspired by insights on the nature of agency and adaptive behavior [1,11,12]. It has been suggested that even if some of these aspects are defined exogenously they could capture general components of all physical systems and could potentially be derived in a more objective manner compared to task based utilities [20]. Moreover these views suggest that the inherent conditions of precariousness and the perturbations an agent must face are crucial ingredients for the emergence of purpose generating mechanisms. In that sense, our main concern has been to explore an instance of the conditions in which a stable set of attracting states arises, conferring value to observations and leading to what seemed as self-sustaining dynamics. Although all measurements lacked any initial functional value, the model presupposes the capacity of the agent to measure its operational integrity as it would occur in an organism monitoring its bodily states. This raises the issue of establishing more principled protocols to define what should constitute the internal milieu of an agent.

Agent-Environment Coupling: A matter of further analysis, also motivated by results in [4], is the role of the environment to provide structure to the behavior of the agent. For instance, in the environments in [4], a distribution of preferences spontaneously built on the initial set of visual observations tends to correlate with good performance on the task. In the work presented here the initial set of internal measurements afforded by the environment contributes to the formation of a steady state, with the visual features informing the actions necessary to maintain it. Hence similarly to [4], the initial conditions of the agent-environment coupling that furnish the distribution $p(v)$ provide a starting solution for the problem of self-maintenance as long as the agent is able to preserve the statistics. Thus if the agent lacks a sophisticated sensory apparatus, the capacity to extract invariances or the initial statistics of sensory data do not favor the emergence of goal-seeking behavior, tracking its internal configuration may suffice for some situations. However this requires further unpacking, not only because as discussed earlier it remains uncertain how to define the internal aspects of an agent, but also because often simulations do not capture the essential characteristics of real environments either [8].

Drive Decomposition: While here we have afforded our model certain levels of independence between the sensory data and the internal measurements, it might be sensible to imagine that internal states would affect perception and perceptual misrepresentation would affect internal states. Moreover, as the agent moves from normative conditions based entirely on viability to acquire other higher level preferences, it learns to integrate and balance different drives and goals. From Eq. 8 it is also possible to conceive a simplified scenario and establish the following expression (Appendix D),

$$\mathbf{G}(\pi, \tau) \approx \underbrace{\mathbb{E}_{q(o_\tau, v_\tau, \theta, s_\tau | \pi)}[\ln q(s_\tau | \pi) - \ln p(s_\tau | o_\tau, \pi)]}_{epistemic\ value}$$

$$- \underbrace{\mathbb{E}_{q(o_\tau, v_\tau, \theta, s_\tau | \pi)}[\ln p(o_\tau)]}_{high\ level\ value}$$

$$+ \underbrace{\mathbb{E}_{q(o_\tau, s_\tau | \pi)} H[p(v_\tau | s_\tau, o_\tau, \pi)]}_{regulatory\ value} \tag{7}$$

Where the goal-seeking value is decomposed into a component that considers preferences encoded in a distribution $p(o)$ and another element estimating the expected entropy of the distribution of essential variables. Policies would balance the contributions resolving for hypothetical situations, such as a higher level goal being at odds with the viability of the system.

Acknowledgment. This research utilised Queen Mary's Apocrita HPC facility, supported by QMUL Research-IT. doi:10.5281/zenodo.438045.

A   Expected Free Energy with Measurements v

We consider a generative model $p(s, o, v|\pi)$ for the EFE equation and obtain a joint distribution of preferences $p(o, v)$. If we are interested exclusively in v, assuming and treating o and v as if they were independent modalities, and ignoring o we obtain:

$$
\begin{aligned}
\mathbf{G}(\pi, \tau) &= \mathbb{E}_{q(o_\tau, v_\tau, \theta, s_\tau|\pi)}[\ln q(s_\tau|\pi) - \ln p(s_\tau, o_\tau, v_\tau|\pi)] \qquad (8)\\
&\approx \mathbb{E}_{q(o_\tau, v_\tau, \theta, s_\tau|\pi)}[\ln q(s_\tau|\pi) - \ln q(s_\tau|, o_\tau, v_\tau, \pi) - \ln p(o_\tau, v_\tau)]\\
&\approx \mathbb{E}_{q(o_\tau, v_\tau, \theta, s_\tau|\pi)}[\ln q(s_\tau|\pi) - \ln q(s_\tau|, o_\tau, v_\tau, \pi) - \ln p(o_\tau) - \ln p_\theta(v_\tau)]\\
&\approx \mathbb{E}_{q(o_\tau, v_\tau, \theta, s_\tau|\pi)}[\ln q(s_\tau|\pi) - \ln q(s_\tau|, o_\tau, v_\tau, \pi) - \ln p_\theta(v_\tau)]\\
&\approx -\mathbb{E}_{q(o_\tau, v_\tau, \theta|\pi)} D_{KL}[q(s_\tau|, o_\tau, v_\tau, \pi)||q(s_\tau|\pi)] - \mathbb{E}_{q(v_\tau, \theta, s_\tau|\pi)}[\ln p_\theta(v_\tau)]\\
&\qquad\qquad\qquad (9)
\end{aligned}
$$

B   Novelty and salience

The derivation is equivalent to those found in the classical tabular descriptions of active inference where instead of learning transitions via a function approximator, a mapping from hidden states to observations is encoded by a likelihood matrix $\mathbf{A}$. In the tabular case the beliefs of the probability of an observation given a state are contained in the parameters a_{ij}, which are updated as the agent obtains a particular observation.

$$
\begin{aligned}
\mathbf{G}(\pi, \tau) &= \mathbb{E}_{q(o_\tau, s_\tau, v_\tau, \phi|\pi)}[\ln q(s_\tau, \phi|\pi) - \ln p(o_\tau, v_\tau, s_\tau, \phi|\pi)]\\
&= \mathbb{E}_{q(o_\tau, s_\tau, v_\tau, \phi|\pi)}[\ln q(\phi) + \ln q(s_\tau|\pi)\\
&\qquad - \ln p(\phi|s_\tau, o_\tau, v_\tau, \pi) - \ln p(s_\tau|o_\tau, v_\tau, \pi) - \ln p(o_\tau, v_\tau)]\\
&\approx \mathbb{E}_{q(o_\tau, s_\tau, v_\tau, \phi|\pi)}[\ln q(\phi) + \ln q(s_\tau|\pi)\\
&\qquad - \ln q(\phi|s_\tau, o_\tau, v_\tau, \pi) - \ln q(s_\tau|o_\tau, v_\tau, \pi) - \ln p_\theta(v_\tau)]\\
&\approx \mathbb{E}_{q(o_\tau, s_\tau, v_\tau, \phi|\pi)}[\ln q(s_\tau|\pi) - \ln q(s_\tau|o_\tau, v_\tau, \pi)]\\
&\qquad + \mathbb{E}_{q(o_\tau, s_\tau, v_\tau \phi|\pi)}[\ln q(\phi) - \ln q(\phi|s_\tau, o_\tau, v_\tau, \pi)]\\
&\qquad - \mathbb{E}_{q(o_\tau, s_\tau, v_\tau, \phi|\pi)}[\ln p(v_\tau)]\\
&\approx -\mathbb{E}_{q(o_\tau, s_\tau, v_\tau, \phi|\pi)}[\ln q(s_\tau|o_\tau, v_\tau, \pi) - \ln q(s_\tau|\pi)]\\
&\qquad - \mathbb{E}_{q(o_\tau, s_\tau, v_\tau, \phi|\pi)}[\ln q(\phi|s_\tau, o_\tau, v_\tau, \pi) - \ln q(\phi)]\\
&\qquad - \mathbb{E}_{q(o_\tau, s_\tau, v_\tau, \phi|\pi)}[\ln p(v_\tau)]\\
&\approx -\underbrace{\mathbb{E}_{q(o_\tau, v_\tau, \phi|\pi)}\big[D_{KL}[q(s_\tau|o_\tau, v_\tau, \pi)||q(s_\tau|\pi)]\big]}_{salience}\\
&\qquad - \underbrace{\mathbb{E}_{q(o_\tau, v_\tau, s_\tau|\pi)}\big[D_{KL}[q(\phi|s_\tau, o_\tau, v_\tau, \pi)||q(\phi)]\big]}_{novelty}\\
&\qquad - \underbrace{\mathbb{E}_{q(o_\tau, v_\tau, s_\tau, \phi|\pi)}[\ln p(v_\tau)]}_{instrumental\ value} \qquad (10)
\end{aligned}
$$

C   Implementation

We tested on the Flappy Bird environment [31]. The environment sends a non-visual vector of features containing:

- the bird y position
- the bird velocity.
- next pipe distance to the bird
- next pipe top y position
- next pipe bottom y position
- next next pipe distance to the bird
- next next pipe top y position
- next next pipe bottom y position

The parameter θ of the Bernoulli distribution $p(v)$ was estimated from a *measurement buffer* (i.e. queue) containing the last N values of v gathered by the agent. We tested the agents with large buffers (e.g.. 20^6) as well as small buffers (e.g.. 20) without significant change in performance. The results reported in Fig. 1 were obtained with small sized buffers as displayed in the hyperparameter table below.

The DQN agent was trained to approximate with a neural network a Q-function $Q_\phi(\{s, \theta\}, .)$. For our case study $s = o$ which contains the vector of features, while θ is the parameter corresponding to the current estimated statistics of $p(v)$. An action is sampled uniformly with probability ϵ otherwise $a_t = \min_a Q_\phi(\{s_t, \theta_t\}, a)$. ϵ decays during training.

For the EFE agent, the transition model $p(s_t | s_{t-1}, \phi, \pi)$ is implemented as a $\mathcal{N}(\{s_t, \theta_t\}; f_\phi(s_{t-1}, \theta_{t-1}, a_{t-1}), f_\phi(s_{t-1}, \theta_{t-1}, a_{t-1}))$. Where a is an action of a current policy π with one-hot encoding and f_ϕ is an ensemble of K neural networks which predicts the next values of s and θ. The surprisal model is also implemented with a neural network and trained to predict directly the surprisal in the future as $f_\xi(s_{t-1}, \theta_{t-1}, a_{t-1}) = -\ln p_{\theta_{t-1}}(v_t)$.

In order to calculate the expected free energy in Eq. 6 from a simulated sequence of future steps, we follow the approach described in appendix G in [33] where they show that an information gain of the form $\mathbb{E}_{q(s|\phi)} D_{KL}[q(\phi|s) || q(\phi)]$ can be decomposed as,

$$\mathbb{E}_{q(s|\phi)} D_{KL}[q(\phi|s) || q(\phi)] = -\mathbb{E}_q(\phi) H[q(s|\phi)] + H[\mathbb{E}_{q(\phi)} q(s|\phi)] \qquad (11)$$

with the first term computed analytically from the ensemble output and the second term approximated with a k-NN estimator [2].

Hyperparameters	DQN	EFE
Measurement v buffer size	20	20
Replay buffer size	10^6	10^6
Batch size	64	50
Learning rate	1^{-3}	1^{-3}
Discount rate	0.99	–
Final ϵ	0.01	–
Seed episodes	5	3
Ensemble size	–	25
Planning horizon	–	15
Number of candidates	–	500
Mutation rate	–	0.5
Shift buffer	–	True

D   Drive decomposition

$$
\begin{aligned}
\mathbf{G}(\pi,\tau) &= \mathbb{E}_{q(o_\tau,v_\tau,\theta,s_\tau|\pi)}[\ln q(s_\tau|\pi) - \ln p(s_\tau,o_\tau,v_\tau|\pi)] \\
&= \mathbb{E}_{q(o_\tau,v_\tau,\theta,s_\tau|\pi)}[\ln q(s_\tau|\pi) - \ln p(v_\tau|s_\tau,o_\tau,\pi) - \ln p(s_\tau,o_\tau|\pi)] \\
&= \mathbb{E}_{q(o_\tau,v_\tau,\theta,s_\tau|\pi)}[\ln q(s_\tau|\pi) - \ln p(s_\tau|o_\tau,\pi) - \ln p(o_\tau) - \ln p(v_\tau|s_\tau,o_\tau,\pi)] \\
&\approx \mathbb{E}_{q(o_\tau,v_\tau,\theta,s_\tau|\pi)}[\ln q(s_\tau|\pi) - \ln p(s_\tau|o_\tau,\pi)] - \mathbb{E}_{q(o_\tau,v_\tau,\theta,s_\tau|\pi)}[\ln p(o_\tau)] \\
&\quad + \mathbb{E}_{q(o_\tau,s_\tau|\pi)}H[p(v_\tau|s_\tau,o_\tau,\pi)]
\end{aligned}
\tag{12}
$$

References

1. Barandiaran, X.E., Paolo, E.D., Rohde, M.: Defining agency: individuality, normativity, asymmetry, and spatio-temporality in action. Adapt. Behav. **17**, 367–386 (2009)
2. Beirlant, J., Dudewicz, E.J., Györfi, L., Dénes, I.: Nonparametric entropy estimation: an overview. Int. J. Math. Stat. Sci. **6**(1), 17–39 (1997)
3. Bellemare, M., Srinivasan, S., Ostrovski, G., Schaul, T., Saxton, D., Munos, R.: Unifying count-based exploration and intrinsic motivation. In: Lee, D.D., Sugiyama, M., Luxburg, U.V., Guyon, I., Garnett, R. (eds.) Advances in Neural Information Processing Systems, vol. 29, pp. 1471–1479. Curran Associates, Inc. (2016)
4. Berseth, G., et al.: SMiRL: Surprise Minimizing RL in Dynamic Environments. arXiv:1912.05510 [cs, stat] (2020)
5. Burda, Y., Edwards, H., Storkey, A., Klimov, O.: Exploration by random network distillation. In: International Conference on Learning Representations (2018)
6. Çatal, O., Nauta, J., Verbelen, T., Simoens, P., Dhoedt, B.: Bayesian policy selection using active inference. arXiv:1904.08149 [cs] (2019)
7. Çatal, O., Wauthier, S., Verbelen, T., De Boom, C., Dhoedt, B.: Deep Active Inference for Autonomous Robot Navigation (2020)

8. Co-Reyes, J.D., Sanjeev, S., Berseth, G., Gupta, A., Levine, S.: Ecological Reinforcement Learning. arXiv:2006.12478 [cs, stat] (2020)
9. Damasio, A.R.: Emotions and feelings: a neurobiological perspective. In: Feelings and Emotions: The Amsterdam Symposium, Studies in Emotion and Social Interaction, pp. 49–57. Cambridge University Press, New York (2004)
10. Dayan, P., Hinton, G.E., Neal, R.M., Zemel, R.S.: The Helmholtz machine. Neural Comput. **7**(5), 889–904 (1995)
11. Di Paolo, E.A.: Organismically-inspired robotics: homeostatic adaptation and teleology beyond the closed sensorimotor loop (2003)
12. Di Paolo, E.A.: Robotics inspired in the organism. Intel **53**(1), 129–162 (2010)
13. Friston, K.: Life as we know it. J. R. Soc. Interface **10**(86), 20130475 (2013)
14. Friston, K., FitzGerald, T., Rigoli, F., Schwartenbeck, P., Pezzulo, G.: Active inference: a process theory. Neural Comput. **29**(1), 1–49 (2016)
15. Friston, K., Kilner, J., Harrison, L.: A free energy principle for the brain. J. Physiol.-Paris **100**(1), 70–87 (2006)
16. Friston, K., Samothrakis, S., Montague, R.: Active inference and agency: optimal control without cost functions. Biol. Cybern. **106**(8), 523–541 (2012). https://doi.org/10.1007/s00422-012-0512-8
17. Friston, K.J., Lin, M., Frith, C.D., Pezzulo, G., Hobson, J.A., Ondobaka, S.: Active inference, curiosity and insight. Neural Comput. **29**(10), 2633–2683 (2017)
18. Hinton, G.E., Zemel, R.S.: Autoencoders, minimum description length and Helmholtz free energy. In: Proceedings of the 6th International Conference on Neural Information Processing Systems, NIPS 1993, pp. 3–10. Morgan Kaufmann Publishers Inc., San Francisco (1993)
19. Jordan, M.I., Ghahramani, Z., Jaakkola, T.S., Saul, L.K.: An introduction to variational methods for graphical models. Mach. Learn. **37**(2), 183–233 (1999). https://doi.org/10.1023/A:1007665907178
20. Kolchinsky, A., Wolpert, D.H.: Semantic information, autonomous agency, and nonequilibrium statistical physics. Interface Focus **8**(6), 20180041 (2018)
21. Millidge, B.: Deep active inference as variational policy gradients. J. Math. Psychol. **96**, 102348 (2020)
22. Mnih, V., et al.: Playing Atari with Deep Reinforcement Learning. arXiv:1312.5602 [cs] (2013)
23. Ostrovski, G., Bellemare, M.G., Oord, A., Munos, R.: Count-based exploration with neural density models. In: International Conference on Machine Learning, pp. 2721–2730. PMLR (2017)
24. Pathak, D., Agrawal, P., Efros, A.A., Darrell, T.: Curiosity-driven exploration by self-supervised prediction. In: Proceedings of the 34th International Conference on Machine Learning, ICML 2017, vol. 70, pp. 2778–2787. JMLR.org, Sydney (2017)
25. Pathak, D., Gandhi, D., Gupta, A.: Self-supervised exploration via disagreement. In: International Conference on Machine Learning, pp. 5062–5071. PMLR (2019)
26. Perez, D., Samothrakis, S., Lucas, S., Rohlfshagen, P.: Rolling horizon evolution versus tree search for navigation in single-player real-time games. In: GECCO 2013, pp. 351–358 (2013)
27. Ramstead, M.J.D., Constant, A., Badcock, P.B., Friston, K.J.: Variational ecology and the physics of sentient systems. Phys. Life Rev. **31**, 188–205 (2019)
28. Ramstead, M.J.D., Badcock, P.B., Friston, K.J.: Answering Schrödinger's question: a free-energy formulation. Phys. Life Rev. **24**, 1–16 (2018)
29. Sajid, N., Ball, P.J., Friston, K.J.: Active inference: Demystified and compared. arXiv:1909.10863 [cs, q-bio] (2020)

30. Sancaktar, C., van Gerven, M., Lanillos, P.: End-to-End Pixel-Based Deep Active Inference for Body Perception and Action. arXiv:2001.05847 [cs, q-bio] (2020)
31. Tasfi, N.: PyGame Learning Environment. Github repository (2016)
32. Tschantz, A., Baltieri, M., Seth, A.K., Buckley, C.L.: Scaling active inference. arXiv:1911.10601 [cs, eess, math, stat] (2019)
33. Tschantz, A., Millidge, B., Seth, A.K., Buckley, C.L.: Reinforcement Learning through Active Inference. arXiv:2002.12636 [cs, eess, math, stat] (2020)
34. Ueltzhöffer, K.: Deep active inference. Biol. Cybern. **112**(6), 547–573 (2018). https://doi.org/10.1007/s00422-018-0785-7

End-Effect Exploration Drive for Effective Motor Learning

Emmanuel Daucé[1,2](✉)

[1] Institut de Neurosciences de la Timone, CNRS/Aix-Marseille Univ,
Marseille, France
emmanuel.dauce@univ-amu.fr
[2] Ecole Centrale de Marseille, Marseille, France

Abstract. Stemming on the idea that a key objective in reinforcement learning is to invert a target distribution of effects, end-effect drives are proposed as an effective way to implement goal-directed motor learning, in the absence of an explicit forward model. An end-effect model relies on a simple statistical recording of the effect of the current policy, here used as a substitute for the more resource-demanding forward models. When combined with a reward structure, it forms the core of a lightweight variational free energy minimization setup. The main difficulty lies in the maintenance of this simplified effect model together with the online update of the policy. When the prior target distribution is uniform, it provides a ways to learn an efficient exploration policy, consistently with the intrinsic curiosity principles. When combined with an extrinsic reward, our approach is finally shown to provide a faster training than traditional off-policy techniques.

Keywords: Reinforcement learning · Intrinsic reward · Model-based exploration · Motor learning

1 Introduction

Recent developments in artificial intelligence have produced important qualitative leaps in the field of pattern recognition, in video games and assisted driving. However, a number of tasks considered as simple, are struggling to find a convincing artificial implementation... This is the field of action selection and motor control. For instance, the fine manipulation of objects, as well as movement in natural environments, and their combination through real time motor control, remain major scientific challenges at present. Compared to the case of video games, reinforcement learning, for instance, remains rather limited in the field of robotic and motor control. The huge improvements "from-scratch" obtained in virtual environments are difficult to transfer to real robotics, where millions of plays can not be engaged under risk of breakage, and simulators are expensive to develop. The brain capability to develop motor skills in a very wide range of areas has thus no equivalent in the field of artificial learning.

T. Verbelen et al. (Eds.): IWAI 2020, CCIS 1326, pp. 114–124, 2020.
https://doi.org/10.1007/978-3-030-64919-7_13

In short, learning a motor command, or a motor skill, happens to become difficult when considering a moderately complex set of effectors, like a multi-joint arm for instance, operating in the physical world. One aspect of the problem is that the set of circuits that process the sensory signals and produce a motor response is composed of digital units, the neurons, operating at fast pace, and adapting rapidly, while, on the other side, the operation space is made of many joints, rigid elements and muscles covering a wide, continuous domain of responses with many degrees of freedom and much longer response times.

The reinforcement learning framework [16] provides a very generic setup to address the question of learning, both from a machine learning and the brain modelling perspectives. It contains many of the interesting constraints that an agent is facing in order to learn a motor act. The theory, however, is constructed around digital (discrete) control principles. The aim of a digital controller is to establish a one to one correspondence between stimuli and actions, by matching the input with a template action. Given a set of pre-defined actions, an agent is expected to pick the one that matches the most the input. In order to learn how to act, the agent is guided by a reward signal, that is much cheaper to extract than an exact set point. Then the choice of the action is monitored by a scalar quantity, the utility, that is the long term sum of rewards [16]. A Reinforcement Learning agent is agnostic about what the world is. It is just acting so as to maximize the utility. Supported by the solid principles of dynamic programming, this agent is expected to end-up in an optimal motor behavior with regards to the reward constraints provided by the environment. A hidden difficulty however lies in finding a good training dataset for the agent. A classical problem in that case is the lack of sampling efficacy due to the sparsity of the rewards, or to the closed-loop structure of the control task, where the examples encountered are statistically dependent on the controller parameters, providing a risk of a self referential loop and local optimum.

The counterpart to digital control is analog control, corresponding to the "classic" way to design a controller in the majority of real-world scenarios. In the analog control case, both the controller and the environment are dynamical systems. The control of a dynamic system generally relies on a *model* [14]. The controller is capable to mimic the behavior of the environment, and know the effect of its actions on the environment. When a certain objective is given, the controller can act so as to reach the objective through *model inversion*, though, in most cases of interest (like the control of an articulated body), the models are not invertible and no single definite control can be established from a given objective [7] without setting up additional and task-specific regularization constraints. This kind of controller needs a forward model, that is generally given, containing a large engineering knowledge about the agent's effector structure and its environment. A problem arises when the learning of a motor command is considered. Such controllers generally lack in adaptivity, and motor adaptation is generally hard to train when the environmental conditions change.

There is thus an apparent trade-off between, on one side, maintaining a model, that may include the many mechanical, environmental and sensory data interactions, and, on the other side, being guided by a simplistic reward, concentrating all the complexity of the external world constraints in a single scalar. The main difference between both approaches appears to be the presence (or the absence) of a definite model of the external world. Knowing the effect of our own actions in the world provides ways to anticipate and do planning to reach an objective, at the cost of maintaining and updating a model of the world [11]. At present time, the trade-off between model-free and model-based control has provoked many debates in the reinforcement learning community, with a preference toward model-free approaches for they are cheaper to maintain and easier to control, leaving unsolved the problem of the sampling efficacy and causing very long training sessions in moderately complex environments.

Our argument here is that the sampling efficacy is bound to the problem of training a model, and one can not expect to have an efficient sampling without a minimal model of the *effect* of action. This model does not need to be perfectly accurate, but it needs to be good enough to allow the agent to efficiently sample its environment in order to grab all the disposable information in relation to the task at hand. We assert here that a simple *effect model* is enough to provide all the needed variability in probing the effect of an action or a policy.

2   Method

2.1   A Probabilistic View to Motor Supervision

The probabilistic view to learning cause-effect relationships is at the core of many recent developments in machine learning, with a body of optimization techniques known as "variational inference" implementing model training from data [10]. We assume for simplicity that the environment is not hidden to the agent, i.e. the environment is fully observable. We also assume a discrete updating of states and actions, like in classic reinforcement learning. Then, if s is the state of the environment (or a context), and a an action performed by the agent, consider e as the *effect* of the action performed by the agent in that particular state.

The effect may reflect, to some point, the result, or the outcome, of the action. Modelling an effect thus supposes that an action should come to an end, a final point, from which it is possible to evaluate or record a result. Consistently with a bunch of neurological observations [5], a simple end-effector open-loop control is here assumed to take place, with a compositional motor command [8] driving a multi-joint effector toward a fixed point, without feedback during the motor realization.

The effect can be a short-term effect, like reading a new state from the environment. It can also be a long term effect, like winning or loosing a game, or reaching an objective s^* in the future. Because there is a lot of uncertainty on the effect of an action, it is modeled as a probability distribution $p(E|s, a)$. When the effect is not dependent on a context, it can be noted more simply $p(E|a)$. Given a certain policy $\pi(a|s)$, one can also consider the average distribution of effects obtained when applying that specific policy, namely

$$p(E|s) = \mathbb{E}_{a \sim \pi(A|s)} p(E|s, a) \tag{1}$$

This marginal distribution is said the effect distribution. By construction, it is dependent on a particular policy (possibly stochastic) over states and actions, and may, for instance, represent the invariant measure of an MDP under that given policy. In our case however, we mostly consider the open-loop control case. The policy is defined over the elements of a compositional action, that is choosing the components of an action. The end-effect of such a compositional action is the terminal state attained at the end of the action, without reading the intermediate states.

In goal-directed control, if e is an expected effect, an *inverse control policy*, whose role is to maximize the chance to reach the effect, can be defined using Bayes rule as:

$$\pi(a|s, e) = \frac{p(e|s, a)\pi(a|s)}{p(e|s)} \tag{2}$$

That is the inversion of the model in a probabilistic setup [1]. Here the marginal effect distribution plays the role of a set point, that fixates the distribution of states toward which the action should head for.

Assume now $p^*(e|s)$ be a target distribution of effects. This distribution is distinct from $p(e|s)$ that is the distribution of effects under the current policy. It is assumed to be realizable from an (unknown) target policy $\pi^*(a|s)$, that can be decomposed into:

$$\pi^*(a|s) = \mathbb{E}_{e \sim p(E|s,a)} \pi^*(a|s, e) \frac{p^*(e|s)}{p(e|s, a)} \tag{3}$$

The right side of the equation provides an estimation of the optimal policy based on a sample e of the effect of the action. Unfortunately, the optimal inverse control policy $\pi^*(a|s, e)$ is unknown. A shortcut is to approximate it with the current inverse control policy $\pi(a|s, e)$. In that case, it happens from Eq. (2) that the formula simplifies to :

$$\pi^*(a|s) \simeq \mathbb{E}_{e \sim p(E|s,a)} \pi(a|s, e) \frac{p^*(e|s)}{p(e|s, a)} \tag{4}$$

$$= \pi(a|s) \mathbb{E}_{e \sim p(E|s,a)} \frac{p^*(e|s)}{p(e|s)} \tag{5}$$

This formula shows that a correction term can be applied to the current policy without owning an explicit forward model, but rather through reading the average effect of the policy. This forms the basis of a supervised approach to motor learning, allowing to update a policy so as to reach a target marginal distribution.

For instance, in a dicrete setup, assuming there exists a $Z(s) \in \mathbb{R}$ such that $\log \pi(a|s) = \beta Q(s, a) + Z(s)$ (softmax policy) makes it possible to update $Q(s, a)$ with the last effect sample e like:

$$Q(s, a) \leftarrow Q(s, a) - \frac{\alpha}{\beta} (\log p(e|s) - \log p^*(e|s)) \tag{6}$$

This update renders the current policy closer to the optimal one. A side effect of this update is that it also changes the effect model that includes the contribution of the policy (see Eq. (1)). Repeating the operation with a small α and different samples of a and e should, on average, reduce the divergence between π and π^*. Equation (6) also provides an interesting identity when considering the classical (reward-based) TD-error, with $\frac{1}{\beta}(\log p(e|s) - \log p^*(e|s))$ taking the role of the TD error, i.e. being identified with $Q(s, a) - \tilde{R}(e)$ (with $\tilde{R}(e)$ a putative sum of rewards up to e), making it possible, for instance, to set up an intrinsic reward implementing a policy that realizes a known prior on the effects. This intrinsic reward is called here the "End-Effect Drive".

This supervised approach to policy relies on an effect model $p(e|s)$ that is less detailed than a forward model. Various kinds of approximate forward models can be found in goal-directed motor control literature, like dynamic goals [9] and distal teacher [7], though generally learning to associate the current action with a distal effect [11,15]. In our case, the model knows nothing about the actions that are performed by the agent. Only the end-effects are recorded to build the model. This "action-agnostic" forward model is close to the concept of state-visit counter, as it is proposed in [2].

2.2   A Uniform Exploration Drive

An important special case is when the objective is not to reach a certain effect, but rather to explore uniformly the range of all possible effects. In that case, the objective effect distribution p^* is uniform over the effect space. This kind of supervision can be seen as a generalization of multiple-goal supervision [6] toward defining each possible effect as a goal. The expected outcome of this uniform drive is to provide a uniform sampling over the effect space, i.e. implement a uniform exploration policy. This intrinsic reward is called here the "End-Effect Exploration Drive" (E3D in short). It is consistent with the pseudo-count bonus proposed in [2]. A similar drive was also proposed in a recent draft as the "state marginal matching" drive [12].

By construction, the E3D is positive when e is rarely visited under the current policy $(p(e|s) < p^*(e))$, and negative the other way. It thus tries to promote rare and "surprising" effects, and lower the occurrence of habitual "boring" effects. It must be noticed that the promotion of rare effects tends to make them less rare, and the rejection of habitual effects tends to make them less habitual, up to an equilibrium where the log-probability ratio should be close to zero. Though the circular dependence between the policy update and the effect model update can provoke some convergence issues, and the equilibrium may not be reached in the case of too fast fluctuations of both distributions during the training process. Some form of regularization is needed in most cases, and, most importantly, should be counterbalanced with some form of utility drive, in order to implement policy optimization through reward maximization. This is the reason why a variational inference setup is particularly well suited in that case, with the distal uniform drive taking the role of a prior under a variational formulation.

2.3 Link with Variational Inference

A key intuition in Friston's "unified brain theory" paper [4] is interpreting the utility, as it is defined in economy and reinforcement learning, as a measure of the negative surprise (i.e the log probability over the sensory data distribution). Combined with a prior distribution in the control space, the action takes the role of a latent variable that is updated so as to reduce the prediction error with regards to the prior, much like in predictive coding.

The unified approach proposed by Friston and colleagues is more generally consistent with the variational auto-encoder principles, in which a latent description of the data is constructed so as to implement a trade-off between the complexity of the description and the accuracy of the prediction. Variational reinforcement learning was recently proposed as a way to reconcile the discrete view to motor control with the continuous processing of the latent variable in variational auto-encoders [3,6,13], with the motor command playing the role of a latent code for the reward data. In our simplified writing, the utility maximization (or surprise minimization) rests on minimizing:

$$- \mathbb{E}_{a \sim \pi(a|s); e \sim p(e|s,a)} \beta R(e) + \mathrm{KL}(\pi(a|s) \| \pi^*(a)) \tag{7}$$

with $R(e)$ here a measure of the sum of (extrinsic) rewards, up to e. Interestingly, the current policy $\pi(a|s)$ lies at the crossroad of a reference (maximum entropy) policy π^* and reward maximization, with the softmax policy representing a compromise between both tendencies in the discrete case.

Extending toward a uniform prior over the space of effects can be written when both considering the motor command and the effect as latent variables that may both explain the current observation, that writes:

$$- \mathbb{E}_{a,e \sim p(a,e|s)} \beta R(e) + \mathrm{KL}(p(a,e|s)) \| p^*(a,e)) \tag{8}$$

For the purpose of illustration, we propose here an additional simplification, that is assuming an *independence* of both factors (e and a) on causing the current data (Naïve Bayes assumption), dividing the KL term in two parts:

$$- \mathbb{E}_{a \sim \pi(a|s); e \sim p(e|s,a)} \beta R(e) + \mathrm{KL}(\pi(a|s)) \| \pi^*(a)) + \mathrm{KL}(p(e|s)) \| p^*(e)) \tag{9}$$

This forms the baseline of our variational policy update setup. The optimization, that is done on $\pi(a|s)$, obeys in that case on a triple constraint, that is maximizing the reward through minimizing both the distance to a baseline policy and the distance of the effect to a reference (supposedly uniform) effect distribution.

In a discrete setup, the uniform prior on the action is supposed implemented with the softmax decision rule. It is then sufficient to assume the following update for the action-value function. After reading e, the TD-error should be defined as:

$$\mathrm{TD}(s,a,e) = \lambda(Q(s,a) - R(e)) + \frac{1}{\beta}(\log p(e|s) - \log p^*(e))) \tag{10}$$

With λ a precision hyperparameter accounting for the different magnitudes of rewards, allowing to manipulate the balance between reward seeking and exploration-seeking drives. Interestingly, the reward $R(e)$ has here the role of a regularizer with regards to the current Q-value. The sum of the future rewards can be estimated using the classical Bellman recurrence equation, i.e. $Q(s,a) \sim r(s,a) + Q(s',a')$, in which case the training procedure needs to maintain an estimate of a standard action-value function $Q_{ref}(s,a)$ to update the actual parameters of the policy $Q(s,a)$.

3   Results

We present in simulation a pilot implementation of the principle presented in the previous section. The principal idea is to illustrate an important feature of biological motor control, that is the control of an effector showing many degrees of freedom, like e.g.. an articulated limb with many joints.

Let $\mathcal{A}$ a control space accounting for a single degree of freedom (here a discrete set of actions i.e. $\mathcal{A} = \{E, S, W, N\}$), each motor command owning n degrees of freedom, i.e. $a_{1:n} = a_1, ..., a_n \in \mathcal{A}^n$. The effect space is expected to be much smaller, like it is the case in end-effector control, where only the final set point of a movement in the peripheral space is considered as the result of the action. Each degree of freedom is supposed to be independent, i.e. the choice of a_i does not depend on the choice of a_j, so that $\pi(a_{1:n}|s_0) = \pi(a_1|s_0) \times ... \times \pi(a_n|s_0)$. When a single context s_0 is considered, the policy writes simply $\pi(a_{1:n})$. The size of the action space is thus combinatorially high, and one can not expect to enumerate every possible action in reasonable computational time. In contrast, the effect space is bounded, and the number of all final states can be enumerated. However, the environment is constructed in such a way that some final states are very unlikely to be reached under a uniform sampling of the action space.

The environment we consider is a grid world with only 18 states and two rooms, with a single transition allowing to pass from room A to room B (see Fig. 1). Starting in the upper left corner of room A, the agent samples a trajectory $a_{1:7} \in \mathcal{A}$ from a policy π, that trajectory being composed of 7 elementary displacements. The agent does not have the capability to read the intermediate states it is passing through, it can only read the final state after the full trajectory is realized. In such an environment, a baseline uniform exploration does not provide a uniform distribution of the final states. In particular, when acting at random, the chance to end-up in the first room is significantly higher than the chance to end up in the second room.

The agent starts from scratch, and has to build a policy $\pi(a)$ and an effect model $p(s_n)$, with s_n the final state. There are two task at hand. A first task is a simple exploration task and the objective is to uniformly sample the effect space, which should imply a non-uniform sampling policy. A second task consists in reaching the lower-right corner, the state that shows the lowest probability with a uniform sampling. For that, a reward of 1 is given when the agent reaches the lower corner, and 0 otherwise.

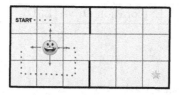

Fig. 1. A simple two-rooms environment. Starting from the upper-left corner, the agent is asked to plan a full sequence made of 7 elementary actions $a_1, ..., a_7$, each elementary action being in (E,S,W,N). The only read-out from the environment is the final state, and a reward, that is equal to 1 if the final state is the lower-right corner, and 0 otherwise.

The update procedure is made of a single loop (see Algorithm 1). The update is done online at each trial. Both the policy and the effect model are updated, with different training parameters. The general idea is to train the effect model a little more "slowly" than the policy, for the policy improvement to firmly take place before they are passed on the effect model.

Stemming from a uniform policy, the effect of the E3D drive is to render the "rare" states more attractive, for they are bound with a positive intrinsic reward, while the more commonly visited states are bound with a negative reward, that reflects a form of "boredom". Marking a rare state as "attractive" tends to increase the number of visits, and finally lower the initial positive reward. In the case of a "gradient" in the likelihood of the final states, with a number of final visits inversely proportional to the distance to the initial state, the E3D drive favors a progressive "expansion" of the visiting territory, for each peripheral state attained will increase the probability to reach its further neighbors, up to the final limit of the state space. In small environment like the one proposed here, the limit is rapidly attained and a rapid alternation of visits is observed over the full state space.

The final distribution of states is compared in Fig. 2 in the case of a uniform policy and the E3D drive. In that specific setup, a strong bias in favor of the first room is observed, and a gradient of likelihood is observed from the initial state toward the lower right corner (Fig. 2A). In contrast, a time consistent uniform pattern of visit is observed in the second case, that illustrates the capability of the E3D drive to set up specific polices devoted to the wide exploration of the environment.

When a reward r is provided by the environment, the question comes whether to balance the policy update procedure in favor of seeking for rewards or seeking for novel effects. By construction, the exploration drive is insensitive to the value of β, for the update is exactly proportional to $\frac{1}{\beta}$. A high β is associate with a small update and vice versa. this is not the case for the leftmost part of the update (Eq. 10). A high β render the agent more sensitive to the extrinsic rewards. In practice, while no reward (or a uniform reward) is provided by the environment, the agent is only guided by the exploration drive. Once a reward is encountered, it tends to overtake the initial uniform exploration,

Algorithm 1. End-Effect Exploration Drive (E3D)

Require: α, β, λ, η

$\quad Q \leftarrow 0_{|\mathcal{A}| \times n}$

$\quad p \leftarrow$ Uniform

$\quad p^* \leftarrow$ Uniform

$\quad$**while** number of trials not exceeded **do**

$\quad\quad$ sample $a_{1:n} \sim \pi(A_{1:n})$

$\quad\quad$ read s_n, r

$\quad\quad p \leftarrow (1 - \eta)p + \eta 1_{S = s_n}$

$\quad\quad$**for** $i \in 1..n$ **do**

$\quad\quad\quad Q(a_i) \leftarrow (1 - \alpha\lambda)Q(a_i) + \alpha\lambda r - \frac{\alpha}{\beta}(\log p(s_n) - \log p^*(s_n))$

$\quad\quad$**end for**

$\quad$**end while**

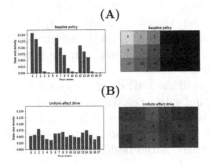

(A)

(B)

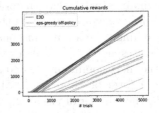

Fig. 2. (a) Task 1 : no reward is provided by the environment. Empirical distribution of the final states, after 5000 trials. **A.** Uniform policy. **B.** End-Effect Exploration Drive (E3D) algorithm. $\alpha = 0.3$, $\beta = 1$, $\lambda = 0.03$, $\eta = 10^{-2}$.

Fig. 3. (b) Task 2 : a reward $r = 1$ is provided by the environment when the agent reaches the lower-right corner. Cumulative sum of rewards over 5000 trials, on 10 training sessions. The E3D algorithm is compared with state-of-the-art epsilon-greedy update. $\alpha = 0.3$, $\beta = 100$, $\lambda = 0.03$, $\eta = 10^{-2}$, $\varepsilon = 0.1$.

providing a firm tendency toward a reward-effective selection of action. This is in contrast with the standard epsilon-greedy strategy, imposing to balance the exploration/exploitation trade-off by hand.

The E3D approach finally provides an Online/on-Policy training procedure that conforms to the main requirements of efficient reinforcement learning, showing both an efficient exploration policy when the rewards are sparse, and the capability to monitor the exploration/exploitation tradeoff with the inverse temperature β in function of the magnitude of the rewards.

The cumulative rewards obtained with the E3D update and a state-of-the-art off-policy/epsilon greedy update are compared in Fig. 3, with ε set to 0.1. If both techniques manage to reach a reward-efficient policy in the long run, the exploration strategy developed in E3D makes it easier to reach the rewarding state, providing the reward earlier in time and developing a fast-paced reward-seeking strategy that overtakes the baseline approaches.

4 Conclusions

Despite its simplicity, our pilot training setup allows to illustrate the main features expected from the inversion of a target distribution of effects, that is the capability to rapidly explore the environment through a reciprocal update of the policy and the effect model. We found in practice that the update of the model needs to be a bit slower than that of the policy to allow for the policy to improve over time and increase the extent of the effect space in a step by step manner. By balancing the effect of the rewards with the inverse temperature parameter, it is possible to catch and exploit very sparse rewards in large environments.

The model is developed here as a first draft in an "end-effect" setup, with very little influence of the context or the states visited in the monitoring of the policy. Extensions toward closed-loop state-action policies is not far from reach, for Eq. (10) allows to exploit the Bellman recurrence to guide the exploration-driven policy with a reference action-value function, that should be updated in parallel to the model and the current policy. Extensions toward continuous action spaces are also needed in order to address effective motor control learning, which resorts to a deeper interpretation of our approach toward the variational inference setup.

References

1. Bays, P.M., Wolpert, D.M.: Computational principles of sensorimotor control that minimize uncertainty and variability. J. Physiol. **578**(2), 387–396 (2007)
2. Bellemare, M., Srinivasan, S., Ostrovski, G., Schaul, T., Saxton, D., Munos, R.: Unifying count-based exploration and intrinsic motivation. In: Advances in Neural Information Processing Systems, pp. 1471–1479 (2016)
3. Fox, R., Pakman, A., Tishby, N.: Taming the noise in reinforcement learning via soft updates. arXiv preprint arXiv:1512.08562 (2015)
4. Friston, K.: The free-energy principle: a unified brain theory? Nat. Rev. Neurosci. **11**(2), 127–138 (2010)
5. Graziano, M.S., Taylor, C.S., Moore, T.: Complex movements evoked by microstimulation of precentral cortex. Neuron **34**(5), 841–851 (2002)
6. Haarnoja, T., Tang, H., Abbeel, P., Levine, S.: Reinforcement learning with deep energy-based policies. arXiv preprint arXiv:1702.08165 (2017)
7. Jordan, M.I., Rumelhart, D.E.: Forward models: supervised learning with a distal teacher. Cogn. Sci. **16**(3), 307–354 (1992)
8. Kadmon Harpaz, N., Ungarish, D., Hatsopoulos, N.G., Flash, T.: Movement decomposition in the primary motor cortex. Cereb. Cortex **29**(4), 1619–1633 (2019)
9. Kaelbling, L.P.: Learning to achieve goals. In: IJCAI, pp. 1094–1099. Citeseer (1993)
10. Kingma, D.P., Welling, M.: Auto-encoding variational Bayes. arXiv preprint arXiv:1312.6114 (2013)

11. Kurutach, T., Clavera, I., Duan, Y., Tamar, A., Abbeel, P.: Model-ensemble trust-region policy optimization. arXiv preprint arXiv:1802.10592 (2018)
12. Lee, L., Eysenbach, B., Parisotto, E., Xing, E., Levine, S., Salakhutdinov, R.: Efficient exploration via state marginal matching. arXiv preprint arXiv:1906.05274 (2019)
13. Levine, S., Koltun, V.: Guided policy search. In: International Conference on Machine Learning, pp. 1–9 (2013)
14. Miall, R.C., Wolpert, D.M.: Forward models for physiological motor control. Neural Netw. **9**(8), 1265–1279 (1996)
15. Mishra, N., Abbeel, P., Mordatch, I.: Prediction and control with temporal segment models. arXiv preprint arXiv:1703.04070 (2017)
16. Sutton, R.S., Barto, A.G.: Reinforcement Learning: An Introduction. MIT Press, Cambridge (2018)

Learning Where to Park

Burak Ergul, Thijs van de Laar, Magnus Koudahl, Martin Roa-Villescas, and Bert de Vries[✉]

Electrical Engineering Department, Eindhoven University of Technology, Groene Loper 19, 5612 AP Eindhoven, The Netherlands
bert.de.vries@tue.nl

Abstract. We consider active inference as a novel approach to the design of synthetic autonomous agents. In order to assess active inference's feasibility for real-world applications, we developed an agent that controls a ground-based robot. The agent contains a generative dynamic model for the robot's position and for performance appraisals by an observer of the robot. Our experiments show that the agent is capable of learning the target parking position from the observer's feedback and robustly steer the robot toward the learned target position.

Keywords: Active inference · Robotics · Variational Bayesian learning

1 Introduction

The idea of autonomously operating synthetic agents is an active research area in the machine learning community. Development of these agents involves a number of hard challenges, for instance the need for agents to be capable of adaptively updating their goals in dynamic real-world settings.

In this project we investigated a novel solution approach to the design of autonomous agents. We recognize that any "intelligent" autonomous agent needs to be minimally capable of realizing three tasks:

- Perception: online tracking of the state of the world.
- Learning: updating its world model in case real-world dynamics are poorly predicted.
- Decision making and control: executing purposeful behavior by taking advantage of its knowledge of the state of the world.

Active Inference (ActInf) is a powerful computational theory of how *biological* agents accomplish the above mentioned task palette. ActInf relies on formulating all tasks (perception, learning and control) as inference tasks in a biased generative model of the agent's sensory inputs [8].

In order to assess the feasibility and capabilities of active inference as a framework for the design of *synthetic* agents in a real-world setting, we develop here an agent for a ground-based robot that learns to navigate to an initially undisclosed location. The agent can only learn where to park through situated interactions with a human observer who is aware of the target location.

© Springer Nature Switzerland AG 2020
T. Verbelen et al. (Eds.): IWAI 2020, CCIS 1326, pp. 125–132, 2020.
https://doi.org/10.1007/978-3-030-64919-7_14

2   Problem Statement

In this design study, we are particularly interested in two issues:

1. Can the agent *learn* the correct target position from situated binary appraisals by a human observer?
2. Can the agent robustly *steer* the robot to the inferred target position?

3   Model Specification

Active inference, a corollary of the free energy principle, brings together perception, learning and control in a unifying theory [8]. Active inference agents comprise a biased generative model that encodes assumptions about the causes of the agent's sensory signals. The generative model is biased in the sense that the agent's goals are encoded as priors over future states or observations.

Following [11,12], the agent's model at time step t in this paper takes the form of a state-space model

$$p_t(o, s, u) \propto p(s_{t-1}) \prod_{k=t}^{t+T} \underbrace{p(o_k|s_k)}_{\text{observation}} \underbrace{p(s_k|s_{k-1}, u_k)}_{\text{state transition}} \underbrace{p(u_k)}_{\text{control}} \underbrace{p'(o_k)}_{\text{goal}}, \qquad (1)$$

where o, s and u refer to the agent's observations, internal states and control signals respectively. Note that the model includes states and observations for T time steps in the future.

The agent's generative model consists of two interacting sub-models: a physical model for the robot's position and orientation and a target model for user appraisals, see Fig. 1. Initially, the physical model has no explicit goal priors. However, the agent's target model infers desired future locations from appraisals and relays this information to the physical model. Thus, as time progresses, the physical model acquires increasingly accurate information about desired future positions.

3.1   The Physical Model

The physical model is responsible for inferring the controls necessary for navigating the agent from any position a to position b. Observations are noisy samples of the robot's position and orientation. The inferred controls are translation and rotation velocities that are used in a differential steering scheme [7].

The states of the physical model are given by $s_k = (x_k, y_k, \phi_k)$ where (x_k, y_k) specify the (latent) position of the agent and ϕ_k the orientation. Controls are given by $u_k = (\Delta\phi_k, \Delta d_k)$, where $\Delta\phi_k$ specifies rotation velocity and Δd_k specifies translation velocity. The transition dynamics are specified as

$$p(s_k|s_{k-1}, u_k) = \mathcal{N}(s_k \,|\, g(s_{k-1}, u_k), 10^{-1}\mathrm{I}) \qquad (2)$$

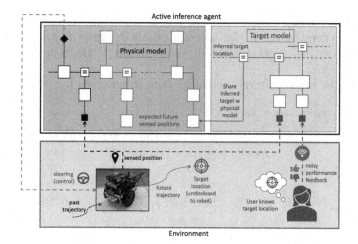

Fig. 1. The information processing architecture of the active inference agent and its environmental interactions. The environment consists of a robot and a human observer that (wirelessly) casts performance appraisals.

where

$$g(s_{k-1}, u_k) = \begin{pmatrix} \phi_{k-1} + \Delta\phi_k \\ x_{k-1} + \Delta d_k \cdot cos(\phi_k) \\ y_{k-1} + \Delta d_k \cdot sin(\phi_k) \end{pmatrix}. \tag{3}$$

In these expressions, $\mathcal{N}(\cdot|m, v)$ is a Normal distribution with mean m and variance v, and I denotes an identity matrix (of appropriate dimension). To couple the observations to internal states, we specify an observation model as

$$p(o_k|s_k) = \mathcal{N}(o_k \mid s_k, 10^{-1}\mathrm{I}) \tag{4}$$

We choose the following controls and state priors:

$$p(u_k) = \mathcal{N}(u_k \mid [0, 0], 10^{-2}\mathrm{I}) \tag{5a}$$
$$p(s_0) = \mathcal{N}(s_0 \mid [0, 0, \pi/2], 10^{-2}\mathrm{I}) \tag{5b}$$

Finally, the goal priors are specified as a prior on observations:

$$p'(o_k) = \mathcal{N}(o_k \mid \hat{o}, 10^{-2}\mathrm{I}). \tag{6}$$

where we denote $\hat{o}$ as a "target parameter" of the physical model.

3.2 The Target Model

The target model is responsible for inferring beliefs about the intended target location $\hat{o}$ by observing user feedback. The inferred beliefs about the target location are subsequently used as a prior belief for the physical model's target parameter $\hat{o}$. The idea of learning a goal prior by a second generative model for additional sensory inputs is further explored in [11]. Technically, the target model is a generative model for user appraisals. In order to reason about the target location, the target model will also be aware of the robot's current and previous position.

Specifically, we use a target model at time step t given by

$$p(r_t, b_t, b_{t-1}, \lambda, \hat{o} \,|\, y_t, y_{t-1}) =$$
$$\underbrace{p(r_t \,|\, b_t, b_{t-1}, \lambda, \hat{o})}_{\text{appraisal}} \cdot \underbrace{p(b_t \,|\, y_t) p(b_{t-1}|y_{t-1})}_{\text{position}} \cdot \underbrace{p(\lambda)}_{\text{precision}} \cdot \underbrace{p(\hat{o})}_{\text{target}} \qquad (7)$$

where

$$p(r_t \,|\, \hat{o}, b_t, b_{t-1}, \lambda) = \text{Bernoulli}(r_t \,|\, \sigma(U(b_t, b_{t-1}, \hat{o}, \lambda))) \qquad (8a)$$
$$p(b_t \,|\, y_t) \qquad\quad = \mathcal{N}(b_t \,|\, y_t, 10\mathrm{I}) \qquad (8b)$$
$$p(b_{t-1} \,|\, y_{t-1}) \qquad = \mathcal{N}(b_{t-1} \,|\, y_{t-1}, 10\mathrm{I}) \qquad (8c)$$
$$p(\lambda) \qquad\qquad\quad = \mathcal{N}(\lambda \,|\, [2, 2], 5\mathrm{I}) \qquad (8d)$$
$$p(\hat{o}) \qquad\qquad\quad = \mathcal{N}(\hat{o} \,|\, o_0, 100\mathrm{I}) \qquad (8e)$$

The model for binary user appraisals uses a "utility" function

$$U(b_t, b_{t-1}, \hat{o}, \lambda) = f(y_t, \hat{o}, \lambda) - f(y_{t-1}, \hat{o}, \lambda) \qquad (9)$$

with

$$f(y, \hat{o}, \lambda) = -\sqrt{(y - \hat{o})^T e^\lambda (y - \hat{o})} \qquad (10)$$

to score the current position y_t to the previous position y_{t-1}, given the current belief over the target $\hat{o}$. λ is a precision parameter governing the width of the utility function. The utility is passed through a sigmoid $\sigma(x) = 1/(1 + e^{-x})$ to parameterize a Bernoulli distribution over binary user appraisals r_t. The user provides appraisals by observing the current and previous positions of the robot. The observed user appraisal is set to 1 if the user thinks that the current robot position is closer to the target than the previous assessment, and otherwise the appraisal is set to 0 (zero). The model was validated in a simulation environment first and later ported to the robot.

The physical model and the target model are linked by drawing a sample from the posterior belief about the intended target location in the target model. This sample is used to parameterize the goal prior of the physical model, i.e., $p'(o_k) = \mathcal{N}(o_k \,|\, \hat{o}^*, 10^{-2}\mathrm{I})$ with $\hat{o}^*$ sampled from $q(\hat{o}|m_{\text{target}})$, see Fig. 2 for the factor graphs of both models.

4   Experimental Validation

4.1   Setup

In this study we design an active inference-based control agent for a two-wheeled robot made by Parallax, Inc. [14]. The actuators of the robot are two continuous-rotation servo motors (one for each wheel) and the robot's sensors include a

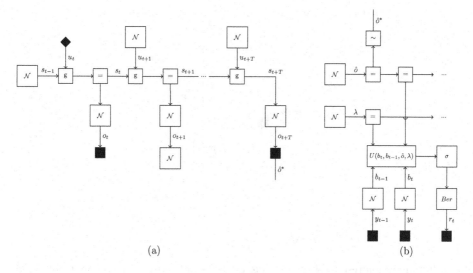

(a)   (b)

Fig. 2. (a) A Forney-style factor graph (FFG) of the physical model. (b) FFG of the Target model. Note that the mean of the future target position in the physical model $(\hat{o}^*)$ is sampled from the posterior belief by the Target model about that position.

gyroscope and two angular position feedback sensors. The agent's control signals are independent (delta) velocity signals to the servo motors. While the gyroscope reports the current orientation of the robot, the angular position feedback sensors are used for determining how many degrees the wheels have rotated. The current position of the robot is calculated by dead reckoning. Dead reckoning is an infrastructure-free localization method where the current position of a mobile entity is calculated by advancing a previously known position using estimated speed over time and course [5].

We employed a Raspberry Pi 4 [9] as a platform for executing free energy minimization (coded in Julia [3], running on Raspberry Pi's Linux variant) and an Arduino Uno [1] for gathering sensor readings and actuating the motors. The Raspberry Pi is wirelessly connected to a PC and user appraisals are provided using this wireless connection.

Inference algorithms were automatically generated using the probabilistic programming toolboxes ForneyLab [4] and Turing [10].

We use an *online* active inference simulation scheme that comprises three phases per time step: (1) act-execute-observe, (2) infer, (3) slide, as described in [15]. The simulation ran for 30 time steps with a horizon $T = 2$.

4.2   Results

Typical simulation results of the trajectory of the robot are shown in Fig. 3. The results show that the agent is capable of steering the robot to the intended target.

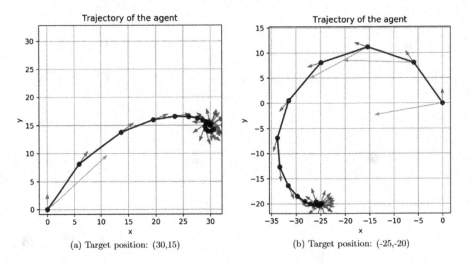

(a) Target position: (30,15)   (b) Target position: (-25,-20)

Fig. 3. Simulation results of the physical model. Green arrows show the orientation of the agent and the red arrows show the proposed motion for the next iteration. (Color figure online)

Figure 4 depicts a typical evolution of the agent's belief about the intended target location. The mean of the belief $\hat{o}$ comes within 2 cm of the target location in approximately 60 iterations.

We also tested the performance of the agent after interventions such as physically changing the orientation of the agent en route. The following video fragment demonstrates how the active inference agent immediately corrects a severe manual interruption and continues its path towards the target location: https://youtu.be/AJevoOmKMO8.

5   Related Work

Prior work on agent-based models within the active inference framework has mainly focused on simulated agents, with a few real-world implementations only recently emerging. In [2] a simulated photo-taxis agent is introduced with a focus on performance evaluation based on achieving goal-directed tasks rather than accurately describing world dynamics. In our work, we followed a similar approach. The physical model introduced in Sect. 3.1 encodes information about world dynamics. A major difference between [2] and this paper is the way goal-directed behavior is induced. In [2] a goal state is not explicitly specified, but rather is a consequence of how priors relating to observations and controls are implemented. In our formulation, a goal state is defined as a prior distribution over future observations.

More recent work, notably [13], addresses the gap between simulated agent implementations and real-world applications. In [13] an active inference model for body perception and actions in a humanoid robot is implemented with a

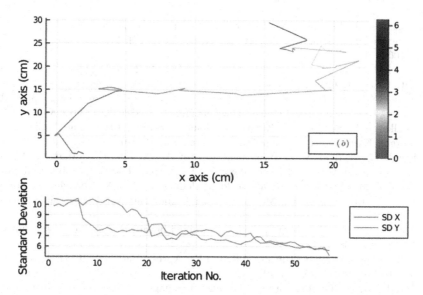

Fig. 4. Simulation results of the target model with a user in the loop. The agent converges to the target location on a 2D plane by observing binary user appraisals. The initial position of the agent is (0,0) and the target location specified by the user is (15,30). The user provides a binary appraisal in each time step.

comparison to classical inverse kinematics. Their results show improved accuracy without an increase in computational complexity providing further evidence for active inference's promise for real-world applications.

6   Conclusions

In order to assess active inference's feasibility for real-world applications, we developed an agent that controls a ground-based robot. The experiments provide support for the notion that active inference is a viable method for constructing synthetic agents that are capable of learning new goals in a dynamic world. More details about this project are available in [6].

Acknowledgements. This work was partly financed by research programmes ZERO and EDL with project numbers P15-06 and P16-25 respectively, which are both (partly) financed by the Netherlands Organisation for Scientific Research (NWO).

References

1. Arduino.cc: Arduino uno rev3. https://store.arduino.cc/arduino-uno-rev3 (2020). Accessed 8 April 2020
2. Baltieri, M., Buckley, C.L.: An active inference implementation of phototaxis. Artif. Life Conf. Proc. **29**, 36–43 (2017). https://doi.org/10.1162/isal_a_011. https://www.mitpressjournals.org/doi/abs/10.1162/isal_a_011

3. Bezanson, J., Edelman, A., Karpinski, S., Shah, V.B.: Julia: a fresh approach to numerical computing. SIAM Rev. **59**(1), 65–98 (2017). https://doi.org/10.1137/141000671
4. Cox, M., van de Laar, T., de Vries, B.: A factor graph approach to automated design of Bayesian signal processing algorithms. Int. J. Approximate Reasoning **104**, 185–204 (2019). https://doi.org/10.1016/j.ijar.2018.11.002. http://www.sciencedirect.com/science/article/pii/S0888613X18304298
5. Edelkamp, S., Schroedl, S., Koenig, S.: Heuristic Search: Theory and Applications. Morgan Kaufmann Publishers Inc., San Francisco (2010)
6. Ergul, B.: A Real-World Implementation of Active Inference. Master's Thesis, Eindhoven University of Technology (2020). https://biaslab.github.io/pdf/msc/Ergul-2020-MSc-thesis-A-Real-World-Implementation-of-Active-Inference.pdf
7. Everett, H.R.: Sensors for Mobile Robots: Theory and Application. A. K. Peters Ltd, Natick (1995)
8. Friston, K.: The free-energy principle: a unified brain theory? Nat. Rev. Neurosci. **11**(2), 127–138 (2010)
9. Gay, W.: Raspberry Pi Hardware Reference. Apress, Berkeley (2014). https://doi.org/10.1007/978-1-4842-0799-4
10. Ge, H., Xu, K., Ghahramani, Z.: Turing: a language for flexible probabilistic inference. In: International Conference on Artificial Intelligence and Statistics, AISTATS 2018, 9–11 April 2018, Playa Blanca, Lanzarote, Canary Islands, Spain, pp. 1682–1690 (2018). http://proceedings.mlr.press/v84/ge18b.html
11. Koudahl, M.T., de Vries, B.: Batman: Bayesian target modelling for active inference. In: ICASSP-2020 Conference, Barcelona (2020)
12. van de Laar, T.W., de Vries, B.: Simulating active inference processes by message passing. Front. Robot. AI **6**, 20 (2019). https://doi.org/10.3389/frobt.2019.00020. https://www.frontiersin.org/article/10.3389/frobt.2019.00020
13. Oliver, G., Lanillos, P., Cheng, G.: Active inference body perception and action for humanoid robots. CoRR abs/1906.03022 http://arxiv.org/abs/1906.03022 (2019)
14. Parallax: Robot shield with arduino. https://www.parallax.com/product/32335 (2020). Accessed 8 April 2020
15. van de Laar, T.: Automated design of Bayesian signal processing algorithms. Ph.D. Thesis, Eindhoven University of Technology (2019)

Active Inference: Theory and Biology

Integrated World Modeling Theory (IWMT) Implemented: Towards Reverse Engineering Consciousness with the Free Energy Principle and Active Inference

Adam Safron[1,2]([⊠])

[1] Kinsey Institute, Indiana University, Bloomington, IN 47405, USA
asafron@gmail.com
[2] Cognitive Science Program, Indiana University, Bloomington, IN 47405, USA

Abstract. Integrated World Modeling Theory (IWMT) is a synthetic model that attempts to unify theories of consciousness within the Free Energy Principle and Active Inference framework, with particular emphasis on Integrated Information Theory (IIT) and Global Neuronal Workspace Theory (GNWT). IWMT further suggests predictive processing in sensory hierarchies may be well-modeled as (folded, sparse, partially disentangled) variational autoencoders, with beliefs discretely updated via the formation of synchronous complexes—as self-organizing harmonic modes (SOHMs)—potentially entailing maximal a posteriori (MAP) estimation via turbo coding. In this account, alpha-synchronized SOHMs across posterior cortices may constitute the kinds of maximal complexes described by IIT, as well as samples (or MAP estimates) from multimodal shared latent space, organized according to egocentric reference frames, entailing phenomenal consciousness as mid-level perceptual inference. When these posterior SOHMs couple with frontal complexes, this may enable various forms of conscious access as a kind of mental act(ive inference), affording higher order cognition/control, including the kinds of attentional/intentional processing and reportability described by GNWT. Across this autoencoding heterarchy, intermediate-level beliefs may be organized into spatiotemporal trajectories by the entorhinal/hippocampal system, so affording episodic memory, counterfactual imaginings, and planning.

"The formal distinction between the FEP and IIT is that the free energy principle is articulated in terms of probabilistic beliefs about some (external) thing, while integrated information theory deals with probability distributions over the states of some system… On the other hand, both the FEP and IIT can be cast in terms of information theory and in particular functionals (e.g., variational free energy and 'phi'). Furthermore, they both rest upon partitions (e.g., Markov blankets that separate internal from external states and complexes that constitute conscious entities and can be distinguished from other entities). This speaks to the possibility of, at least, numerical analyses that show that minimising variational free energy maximises 'phi' and vice versa… This supports the (speculative) hypothesis that adding further constraints on generative models—entailed by systems possessing a Markov blanket—might enable us to say which systems are conscious, and which are not." - Friston et al. [1], Sentience and the Origins of Consciousness: From Cartesian Duality to Markovian Monism.

© Springer Nature Switzerland AG 2020
T. Verbelen et al. (Eds.): IWAI 2020, CCIS 1326, pp. 135–155, 2020.
https://doi.org/10.1007/978-3-030-64919-7_15

Keywords: World models · Global workspaces · Integrated information · Autoencoders · Turbo codes · Phenomenal consciousness · Conscious access

1 Integrated World Modeling Theory (IWMT) Summarized: Combining the Free Energy Principle and Active Inference (FEP-AI) Framework with Integrated Information Theory (IIT) and Global Neuronal Workspace Theory (GNWT)

The Hard problem of consciousness asks, how can it be that there is "something that it is like" to be a physical system [2]? This is usually distinguished from the "easy problems" of addressing why different biophysical and computational phenomena correspond to different qualities of experience. IWMT attempts to address consciousness' enduring problems with the Free Energy Principle [3] and Active Inference [4] (FEP-AI) framework. FEP-AI begins with the understanding that persisting systems must regulate environmental exchanges and prevent entropic accumulation (cf. Good Regulator Theorem from cybernetics) [5]. In this view, minds and brains are predictive controllers for autonomous systems, where action-driven perception is realized via probabilistic inference. FEP-AI has been used to address consciousness in multiple ways [1, 6], with IWMT representing one such attempt. Below I briefly summarize the major claims of IWMT via modified excerpts from the original publication of the theory in *Frontiers in Artificial Intelligence* [7], as well as the accompanying preprint, "IWMT Revisited" [8]. Please see these longer works for further discussion.

IWMT's primary claims are as follows *(originally published in [7])*:

1. Basic phenomenal consciousness is what it is like to be the functioning of a probabilistic generative model for the sensorium of an embodied–embedded agent [9].
2. Higher order and access consciousness are made possible when this information can be integrated into a world model with spatial, temporal, and causal coherence. Here, coherence is broadly understood as sufficient consistency to enable functional closure and semiotics/sense-making [10–12]. That is, for there to be the experience of a world, the things that constitute that world must be able to be situated and contrasted with other things in some kind of space, with relative changes constituting time, and with regularities of change constituting cause. These may also be preconditions for basic phenomenality, especially if consciousness (as subjectivity) requires an experiencing subject with a particular point of view on the world [13–15].
3. Conscious access, or awareness/knowledge of experience—and possibly phenomenal consciousness—likely requires generative processes capable of counterfactual modeling with respect to selfhood and self-generated actions [16, 17].

IIT begins with considering the preconditions for systems to exist intrinsically from their own perspectives, as is observed with the privately-experienced 1st person ontology of consciousness as subjectivity [18]. IIT speaks to the Hard problem by grounding itself in phenomenological axioms, and then goes on to postulate mechanisms that could realize such properties. While IWMT focuses on explaining the functional, algorithmic, and implementational properties that may give rise to consciousness—or experience as a subjective point of view—it also considers ways in which FEP-AI and IIT can be combined as general systems theories and models of causal emergence [19–21]. In brief, IWMT argues that complexes of integrated information (as irreducible self-cause-effect power) are also Markov-blanket-bound networks of effective connectivity associated with high marginal likelihoods and capacity for "self-evidencing" [22].

GNWT has a more restricted scope than IIT and FEP-AI, instead focusing on the properties of computational systems that could realize the functions of consciousness as a means of globally integrating and broadcasting information from otherwise disconnected mental systems [23]. GNWT suggests that workspaces help to select particular interpretations of events, potentially understandable as Bayesian model selection [23, 24], which is highly compatible with IWMT. However, IWMT also potentially differs from the theories it attempts to combine, suggesting that complexes of integrated information and global workspaces only entail subjective experience when applied to systems capable of functioning as Bayesian belief networks and cybernetic controllers for embodied agents [25]. That is, IWMT argues that integration and widespread availability of information are necessary, but not sufficient, preconditions for enabling consciousness. Specifically, *IWMT claims that consciousness is what integrated world-modeling is like, when generative processes are capable of jointly integrating information into models with coherence with respect to space, time, and cause for systems and their relationships with their environments.* These coherences are stipulated to be required for situating modeled entities relative to each other with specific properties, without which there would be no means of generating an experienceable world. IWMT further introduces a mechanism for generating complexes of integrated information and global workspaces via (Markov-blanket-bound) meta-stable synchronous complexes— or "self-organizing harmonic modes" (SOHMs)—wherein synchrony both emerges from and facilitates the integration of information via "communication-through-coherence" [26, 27]. IWMT further suggests that the stream of experience (Fig. 1) is constituted by a series of SOHM-formation events, computationally understood as entailing loopy belief propagation, so generating joint posterior distributions (or maximal estimates derived thereof) over sensoriums of embodied agents as they engage with the environments in which they are embedded.

While parallels can be identified between GNWT and IIT, present discussions and ongoing adversarial collaborations emphasize their differences, such as IIT's claim that consciousness is primarily located in a "posterior hot zone" [28], and GNWT's claim that consciousness requires frontal-lobe engagement for realizing "global availability" of information. IWMT considers both positions to be accurate, but with respect to phenomenal consciousness and conscious access (and other higher-order forms of conscious experience), respectively. That is, frontal cortices are likely required for mental processes such as manipulating and reporting on the contents of consciousness, and so modifying these phenomena in qualitatively/functionally important ways. However, the stream of experience itself may always be generated within hierarchies centered on the posterior cortices, as described in greater detail below.

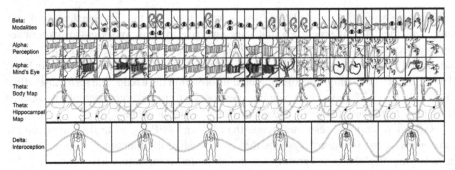

Fig. 1. Depiction of experience with components mapped onto EEG frequency bands.

2 Integrated World Modeling Theory (IWMT) Implemented

2.1 Mechanisms of Predictive Processing: Folded Variational Autoencoders (VAEs) and Self-organizing Harmonic Modes (SOHMs)

IWMT understands cortex using principles from predictive coding [29–31], specifically viewing cortical hierarchies as analogous to VAEs [32, 33], where encoding and generative decoding networks have been folded over at their reduced-dimensionality bottlenecks such that corresponding hierarchical levels are aligned. In this view, hierarchies of superficial pyramidal neurons constitute encoding networks, whose bottom-up observations would be continually suppressed (or "explained away") by predictions from hierarchies of deep pyramidal neurons (and thalamic relays) [34, 35], with only prediction-errors being passed upwards. This is similar to other recent proposals [16], except beliefs are specifically communicated and updated via synchronous dynamics, wherein prediction-errors may be quantized via fast gamma-synchronized complexes [36, 37], and where predictions may take the form of a nested hierarchy of more slowly evolving synchronization manifolds, so affording hierarchical modeling of spatiotemporal events in the world [38] (Fig. 1). More specifically, self-organizing harmonic modes (SOHMs) are suggested to implement loopy belief propagation for approximate inference (cf. turbo coding) [39–41], as well as marginalization over synchronized sub-networks, so instantiating marginal message passing regimes [42].

Combined with mechanisms of divisive normalization and spike-timing dependent plasticity [43–45], this predictive coding setup should induce increasingly sparse connectivity with experience, with all of the functional benefits sparsity provides [46]. These mechanisms (and entailed algorithms) may converge on near-optimal training protocols [47]. With respect to suggestions that the brain may indirectly realize backprop-like computations [48] (Appendix 2), these models view cortical hierarchies as "stacked autoencoders" [49], as opposed to being constituted by a single (folded) VAE. These interpretations of neural computation may be non-mutually exclusive, depending on the granularity with which relevant phenomena evolve. That is, we could think of separate VAEs for each cortical region (e.g. V1, V2, V4, IT), or each cortical macrocolumn [50, 51], and perhaps even each cortical minicolumn. Depending on the timescales over which we are evaluating the system, we might coarse-grain differently [19], with consciousness representing a single joint belief at the broadest level of integration, with the perceptual heterarchy considered as a single VAE. However, a more fine-grained analysis might allow for further factorization, where component subnetworks could be viewed as entailing separate VAEs, with separate Bayesian beliefs. In this view, the cortical heterarchy could be viewed as a single VAE (composed of nested VAEs), as well as a single autoregressive model, where latent beliefs between various VAEs are bound together via synchronous activity, potentially entailing normalizing flows across coupled latent space dynamics [52, 53].

2.2 A Model of Episodic Memory and Imagination

With respect to consciousness, SOHM-formation involving deep pyramidal neurons is suggested to correspond to both "ignition" events as described by GNWT [54], as well as implementation of semi-stochastic sampling from the latent space of VAEs (cf. the "reparameterization trick") [55], including via latent (work)spaces shared by multiple VAEs. If these samples are sequentially orchestrated according to spatiotemporal trajectories of the entorhinal/hippocampal system [56], this may generate a coherent stream of experience. However, coherent sequence transitions between quale states may also potentially be realizable even in individuals without functioning medial temporal lobes, if prior histories of experience allow frontal lobes to enable coherent action-selection and action-driven perception—including with respect to mental acts—in which posterior dynamics may be driven either through overt enaction or via efference copies accompanying covert partial deployment of "forward models" [25].

In this view of the brain in terms of machine learning architectures, the hippocampal complex could be thought of as the top of the cortical heterarchy [57, 58] and spatiotemporally-organized memory register [59]. IWMT suggests this spatial and temporal organization may be essential for coherent world modeling. With respect to grid/place cells of the entorhinal/hippocampal system [60], this organization appears to take the form of 2D trajectories through space, wherein organisms situate themselves according to a kind of simultaneous localization and mapping via Kalman filtering [61]. Anatomically speaking, this dynamic (and volatile) memory system has particularly strong bi-directional linkages with deeper portions of cortical generative models (i.e., reduced-dimensionality latent feature spaces), so being capable of both storing information and shaping activity for these core auto-associative networks. Because of the predictive coding setup—and biasing via neuromodulatory value signals [62, 63]—only maximally informative, novel, unexplained observations will tend to be stored in this spatiotemporally organized memory register. Indeed, this may be one of the primary functions of the hippocampus: temporarily storing information that could not be predicted elsewhere, and then using replay to train relevant subnetworks to be more successfully predictive of likely observations.

As the hippocampus—and cortical systems with which it couples via "big loop recurrence" [64, 65]—re-instantiates trajectories of the organism through space, pointers to prediction errors will be sequentially activated, with the generative model inferring a more complete sensorium based on its training from a lifetime of experience. Computationally speaking, this setup would correspond to a Kalman variational auto-encoder [66]. Experientially speaking, this integration of organismic spatiotemporal trajectories with auto-associative filling-in could provide not just a basis for forming episodic memories, but also the imagination of novel scenarios [67, 68]. Importantly, memories and imaginings can be generated by cortex on its own—given a lifetime of experience with a functioning entorhinal/hippocampal system—but medial temporal lobe involvement appears to be required for these dynamics to be shaped in novel directions that break free of past experience [69–72]. The hippocampal system may further allow for contrasting of anticipated and present estimated states in the process of orchestrating goal-oriented behavior [25, 73] (Appendix 1).

2.3   Brains as Hybrid Machine Learning Architectures

Figure 2 provides a depiction of the human brain in terms of phenomenological correspondences, as well as Marr's computational (or functional), algorithmic, and implementational levels of analysis [74]. On the computational level, various brain functions are identified according to their particular modal character, either with respect to perception (both unconscious and conscious) or action (both unconscious and potentially conscious, via perceptual generative models). On the algorithmic level, these functions are mapped onto variants of machine learning architectures—e.g. autoencoders and generative adversarial networks (Appendix 2), graph neural networks, recurrent reservoirs and liquid state machines—organized according to their potential realization by various systems in the brain. On the implementational level, realizations of algorithmic processes are depicted as corresponding to flows of activity and interactions between neuronal populations, canalized by the formation of SOHMs as metastable synchronous complexes. While the language of predictive processing is used here to help provide bridges to the algorithmic level, descriptions such as vector/tensor fields and attracting manifolds could have alternatively been used in order to remain agnostic as to which algorithms may be entailed by physical dynamics.

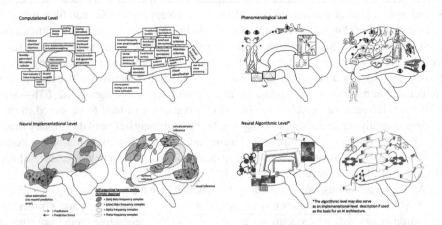

Fig. 2. Depiction of the human brain in terms of phenomenological correspondences, as well as computational (or functional), algorithmic, and implementational levels of analysis.

A phenomenological level is specified to provide mappings between consciousness and these complementary/supervenient levels of analysis. These modal depictions are meant to connotate the inherently embodied nature of experience, but not all images are meant to correspond to the generation of consciousness. That is, it may be the case that consciousness is solely generated by posterior hierarchies centered on the precuneus, lateral parietal cortices, and temporoparietal junction (TPJ) as respective visuospatial (cf. consciousness as projective geometric modeling) [13, 14], somatic (cf. grounded cognition and intermediate level theory) [75–77], and attentional/intentional phenomenology (cf. Attention Schema Theory) [78].

Graph neural networks (GNNs) are identified as a potentially important machine learning architectural principle [79], largely due to their efficiency in emulating physical processes [80–82], and also because the message passing protocols during training and inference may have correspondences with loopy belief propagation and turbo codes suggested by IWMT. Further, grid graphs—potentially hexagonally organized, possibly corresponding to cortical macrocolumns [50], with nested microcolumns that may also be organized as hexagonal grid GNNs, or "capsule networks") [83] —are adduced for areas contributing to quasi-Cartesian spatial modeling (and potentially experience) [84, 85], including the posterior medial cortices, dorsomedial and ventromedial prefrontal cortices, and the hippocampal complex. With respect to AI systems, such representations could be used to implement not just modeling of external spaces, but of consciousness as internal space (or blackboard), which could potentially be leveraged for reasoning processes with correspondences to category theory, analogy making via structured representations, and possibly causal inference.

Neuroimaging evidence suggests these grids may be dynamically coupled in various ways [68], with these aspects of higher-order cognition being understood as a kind of generalized navigation/search process [86, 87]. A further GNN is speculatively adduced in parietal cortices as a mesh grid placed on top of a transformed representation of the primary sensorimotor homunculus (cf. body schemas for the sake of efficient motor control/inference), which is here suggested to have some correspondence/scaling to the body as felt from within, but which may potentially be further morphed to better correspond with externally viewed embodiments (potentially both resulting from and enabling "mirroring" with the bodies of other agents for the sake of coordination and inference) [88]. This partial translation into an allocentric coordinate system is suggested to provide more effective couplings (or information-sharing) with semi-topographically organized representations in posterior medial cortices. The TPJ is depicted as containing a ring-shaped GNN to reflect a further level of abstraction and hierarchical control over action-oriented body schemas—which may influence more somatic-like geometries—functionally entailing vectors/tensors over attentional/intentional processes [89].

Frontal homologues to posterior GNNs are also depicted, which may provide a variety of higher-order modeling abilities, including epistemic access for extended/distributed self-processes and intentional control mechanisms. These higher-order functionalities may be achieved via frontal cortices being more capable of temporally-extended generative modeling [90], and also potentially by virtue of being located further from primary sensory cortices, so affording ("counterfactually rich") dynamics that are more decoupled from immediate sensorimotor contingencies. Further, these frontal control hierarchies afford multi-scale goal-oriented behavior via bidirectional effective connectivity with the basal ganglia (i.e., winner-take-all dynamics and facilitation of sequential operations) and canalization via diffuse neuromodulator nuclei of the brainstem (i.e., implicit policies and value signals) [91–95]. Finally, the frontal pole is described as a highly non-linear recurrent system capable of shaping overall activity via bifurcating capacities [96, 97] —with potentially astronomical combinatorics—providing sources of novelty and rapid adaptation via situation-specific attractor dynamics.

While the modal character of prefrontal computation is depicted at the phenomenological level of analysis, IWMT proposes frontal cortices might only indirectly contribute to consciousness via influencing dynamics in posterior cortices [8]. Speculatively, functional analogues for ring-shaped GNN salience/relevance maps may potentially be found in the central complexes of insects and the tectums of all vertebrates [98], although it is unclear whether those structures would be associated with any kind of subjective experience. Even more speculatively, if these functional mappings were realized in a human-mimetic, neuromorphic AI, then it may have both flexible general intelligence and consciousness. In this way, this figure can be considered to be a sort of pseudocode for potentially conscious (partially human-interpretable) AGI with "System 2" capacities [99–101].

2.4 Conclusion: Functions of Basic Phenomenal Consciousness?

According to IWMT, whenever we have self-organizing harmonic modes (SOHMs), then we also have entailed joint marginal probability distributions (where synchrony selects or discretely updates Bayesian beliefs), some of which may entail consciousness. Functionally speaking, potentially experience-entailing SOHMs—as Markov-blanket-bound subnetworks of effective connectivity and complexes with high integrated information, functioning as workspaces—over coupled visuospatial, attentional/intentional, and somatic hierarchies could provide holistic discriminations between different classes of events in ways that would greatly facilitate coherent action selection and credit assignment. That is, a series of coherently estimated system-world states (even without higher-order awareness or explicit/reflexive knowledge) would be extremely adaptive if it could generate these joint posteriors (or MAP estimates derived thereof) on timescales allowing this information to shape (and be shaped by) action-perception cycles. Since there should be substantial auto-associative linkages across visuospatial, attentional/intentional, and somatic modalities, then the consistency of this mutual information may accelerate the formation of SOHMs, such that beliefs can be updated quickly and coherently enough to have actual organismic-semiotic content (i.e., relevance for the organism and its environment). Further, reentrant signaling across different sources of data may provide a) inferential synergy via knowledge fusion from combining modalities, b) enhanced transfer learning and representational invariance via perspectival diversity (i.e., flexible representation from multiple modalities), and c) sensitivity to higher-order relational information, potentially including causal and contextual factors identified by comparing and contrasting constancies/inconstancies across modalities [102]. Even more, these sources of (mutual) information have natural correspondences with subjectivity in terms of providing a particular point of view on a world, centered on the experience of having/being a body.

Thus, when we identify the kinds of information that could enable adaptive functional synergy in 'processing' sensory data, it becomes somewhat less surprising that there might be "something that it is like." However, such inferential dynamics might require a multi-level hierarchy, with a higher (or deeper) inner-loop capable of iteratively forming and vitiating attracting states [103], so instantiating a kind of "dual phase evolution" [104]. A shallow hierarchy might be overly enslaved to immediate environmental couplings/contingencies [105], and would potentially constitute unconscious inference, with consciousness-entailing states never being generated on any level of abstraction. However, the precise functional boundaries of phenomenal consciousness remain unclear, and is a direction for future work for IWMT.

3 Appendices

3.1 Appendix 1: A Model of Goal-Oriented Behavior with Hippocampal Orchestration

Figure 3 depicts memory and planning (as inference) via predictive processing, orchestrated via the spatiotemporal trajectories of the entorhinal/hippocampal system. In this model, precision-weighting/gain-amplification takes place via "big loop recurrence" with the frontal lobes [64], with the more specific suggestion that selection/biasing of policies over forward models cause efference copies to be projected to posterior generative models. In line with recent proposals [106], the hippocampus can operate with either "predictive-suppressive" or "fictive prediction error" modes, which are here suggested to correspond to degree of coupling with respective posterior vs. frontal cortices, with the former corresponding to direct suppression of observations, and the latter facilitating the 'reinstatement' of memories, and novel imaginings for the sake of planning and causal reasoning [68, 107]. This frontal coupling is hypothesized to be a source of "successor representations" (i.e., population vectors forming predictive anticipatory sweeps of where the organism is likely to go next) via integration of likely policies and action models (via dorsal prefrontal cortex) and evaluations of likely outcomes (via ventral prefrontal cortex).

In this model of generalized navigation, the hippocampal system iteratively contrasts predictive representations with (either sensory-coupled or imaginative) present state-estimates (from coupling with posterior cortices), where prediction-errors both modify future paths, and also allow for encoding of novel information within likely (generalized) spatiotemporal trajectories, given the meta-prior (or inductive bias) that organisms are likely to be pursuing valued goals as they navigate/forage-through physical and conceptual spaces. This alternation may occur at different phases of theta oscillations [108], so affording iterative contrasting of desired and current-estimated states, so canalizing neural activity for goal realization [109], potentially including the formation of complex action sequences (either physical or virtual) via (conscious) back-chaining from potential desired states (i.e., goals) to presently-inferred realities [25]. Theoretically, this kind

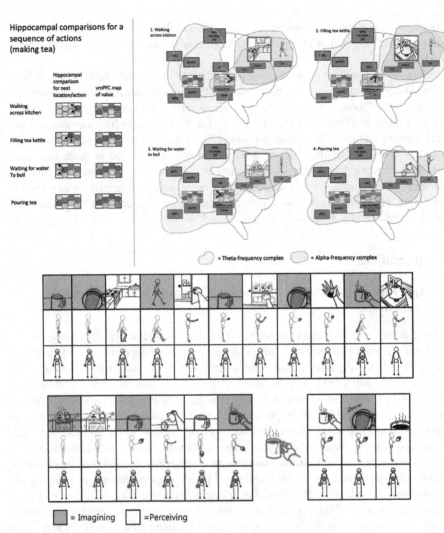

Fig. 3. Hippocampally-orchestrated imaginative planning and action selection via generalized navigation/search.

of iterated contrasting mechanism may also provide a source of high-level analogical (and potentially causal) reasoning [107, 110–112].

By orchestrating alternating counterfactual simulations [73], the hippocampal system may allow for evaluation of possible futures, biased on a moment-to-moment basis by integrated (spatialized) value representations from ventromedial prefrontal cortex [113], and also via "as-if-body loops" with interoceptive hierarchies [114, 115]. In this view of thought as generalized navigation, alternating exploration/sampling of counterfactuals could also be understood as implementing Markov chain Monte Carlo tree search over policy/value space for planning [109, 116]. Theoretically, similar processes could be involved in generating particular actions, if visualization of acts is accompanied by a critical mass of model-evidence (as recurrent activity) accumulating in interoceptive/salience hierarchies [117]. Speculatively, this threshold crossing (or phase transition) may represent a source of readiness potentials [118–120], potentially understood as a kind of "ignition" event and driver of workspace dynamics (understood as high-level Bayesian model selection), corresponding to explosive percolation, triggered by the accumulation of recurrent-activity/model-evidence within upper levels of frontoparietal control hierarchies for enacting/inferring a particular (virtual or physical) proprioceptive state, or pose [121].

3.2 Appendix 2: The VAE-GAN Brain?

Gershman [122] has presented an intriguing account of neural functioning in terms of a powerful class of generative models known as generative adversarial networks (GANs). GANs have many similar use cases to variational-encoders (VAEs) and can even be used in combination for enhanced training as in the case of VAE-GANs [123]. In Gershman's proposal, sensory cortices act as generators which are trained via Turing learning with frontal cortex, which functions as a discriminator and source of (higher-order) consciousness.

IWMT, in contrast, suggests that predictive coding can be understood as generating (basic phenomenal) consciousness via folded VAEs in the ways described above. From this perspective, the ascending and descending streams for each modality constitute respective encoding and generative decoding networks. This is not necessarily inconsistent with Gershman's proposal, in that a sensory hierarchy as a whole can be viewed as a generative network, which relative to the entire brain may provide a VAE-GAN setup. Alternatively, the ascending stream could be interpreted as acting as a discriminator in the GAN sense, in that it is attempting to evaluate the degree to which the descending stream generates veridical images. In this view, folded autoencoders might also be understood as folded GANs, but with folds taking place at output layers of generative decoders and input layers of discriminative encoders. The ascending stream is well-poised to serve this kind of discriminative function in terms of being more directly in touch with the ground truth of sensation and the generative processes of the world, which are the ultimate referents and selection criteria for neural dynamics. This is somewhat different from Gershman's proposal, in that consciousness (as experience) would correspond to generative processes in posterior sensory areas (including multi-modal association cortices), trained via embodied-embedded interaction with the world, with frontal cortices functioning as an elaboration of the generative process in multiple ways, including conscious access via the stabilization and alteration of dynamics within posterior networks, and also via simulated actions and inter-temporal modeling [124].

However, frontal cortex could also be viewed as serving a discriminator function in terms of attentional biasing based on the reliability of information (i.e., precision weighting), mechanistically achieved by altering the gain on excitatory activity from ascending ('discriminative') encoding networks. Thus, frontal cortex could provide a discriminatory function via tuning the sensitivity of ascending perceptual streams. Other non-mutually exclusive possibilities could also be envisioned for a discriminator-like role for frontal cortices:

1. Comparison with memory: "Is what I am perceiving consistent with what I have previously experienced?" This might be particularly important early in development for teaching patterns of attention that promote perceptual coherence.
2. Comparison with causal reasoning: "Is what I am perceiving consistent with what is plausible?" This is closer to Gershman's proposal, wherein failing to establish this discriminatory capacity could increase the probability of delusions and possibly lowered hallucination thresholds in some conditions.
3. Comparison with goal-attainment (a combination of 1 and 2): "Is what I am perceiving consistent with my normal predictions of working towards valued goals?" This could have the effect of adaptively shaping conscious states in alignment with personal (and ultimately organismic) value. According to FEP-AI, all discriminator-like functionality may represent special cases of this highest-level objective: to reduce uncertainty (or accumulate model evidence) with respect to organismic value via perceptual and active inference. Mechanistically, cingulate cortices may have the greatest contributions to generating discrimination signals with respect to overall value [125–130], both through the integrative properties of the cingulum bundle [131, 132], as well as via close couplings with allostatic-organismic interoceptive insular hierarchies [133].

In all of these cases, frontal cortices (broadly construed to include the anterior cingulate) could be viewed as being in an adversarial (but ultimately cooperative) relationship with sensory hierarchies, whose recognition densities would optimize for minimizing perceptual prediction error (i.e., what is likely to be, given data), and where frontally-informed generative densities would optimize for future-oriented (counterfactual) adaptive policy selection (i.e., what ought to be, given prior preferences). In these ways, action and perceptual hierarchies would compete with respect to the ongoing minimization of free energy, while at the same time being completely interdependent for overall adaptive functioning, with both competitive and cooperative dynamics being required for adaptively navigating the world via action-perception cycles. An interesting hybrid of competitive and cooperative dynamics may be found in "learning to learn" via creative imagination and play (including self-play), in which learners may specifically try to maximize surprise/information-gain [134].

With respect to frontal predictions, these may be productively viewed with a 3-fold factorization:

1. A ventral portion representing affectively-weighted sensory outcomes associated with various actions.
2. A dorsal portion representing forward models for enacting sequences that bring about desirable outcomes.
3. A recurrent anterior pole portion that mediates between affect and action selection via its evolving/bifurcating/non-linear attractor dynamics [96, 135, 136].

(1) and (2) would be frontal analogues to the "what" and "where" pathways for vision [90, 137]—with macroscale connectivity reflecting these functional relationships—except here we are dealing with (1) what-where (via coupling with the hippocampal complex) and (2) how-where (via coupling with the parietal lobes). Taken

together (which is how these systems are likely to work under most circumstances), these different parts of frontal cortices could all be understood in a unified sense as implementing policy selection via predictions and precision-weighting.

In Gershman's proposal, he further suggests that predictive coding can be viewed as an efficient way of passing predictions up the cortical hierarchy while removing redundant information. This is consistent with proposals in which the descending stream is interpreted as constituting a means for communicating the backwards propagation of error signals to apical dendrites in cortical layer 1 [138]. Although this (potentially insightfully) inverts the way predictive coding is normally understood, with prediction errors being communicated via the ascending stream, these accounts could potentially be reconciled if we understand perception as involving a circular-causal process of iterative Bayesian model selection. When we consider the capacity for looping effects in networks on the scale of nervous systems—for even the largest deep learning systems, the number of parameters is dwarfed (for now) by those found in a cubic centimeter of cortex—with potentially multiple levels of qualitatively different 'beliefs' (e.g. unconscious sensorimotor, conscious embodiment, and implicit schemas), then it can be difficult to straightforwardly interpret the flow of inference in terms of a clear distinction between predictions and prediction errors. Indeed, hierarchical predictive processing can be viewed as converging on optimal backprop-like functionality via proposals such as "target propagation" and "natural gradient descent" [47, 48]. However, we would also do well to not be overly ecumenical with respect to this potential reconciliation, as more classical accounts of predictive coding induce sparsity on multiple levels, so creating many highly desirable computational properties such as energy efficiency, robustness, and sensitivity to coincidence detection [46, 139]. As such, framing the descending stream as a backpropagation signal may be an account that is both misleading and impoverished with respect to biological realities.

In terms of the potential complexity of cortical generative models, we may want to think of at least three coupled systems that are ultimately integrated as parts of a unified control hierarchy, but which can temporally evolve independently:

1. Unconscious/preconscious lower-level sensorimotor hierarchies with fast fine-grained dynamics for coupling with the environment [96].
2. Conscious mid-level sensorimotor representations with more coarse-grained spatial and temporal dynamics [75].
3. Higher-level abstract re-representations over recognition and generative densities, with unconscious/preconscious dynamics [140], and which may bidirectionally couple with lower and middle levels.

In this way, we could potentially dissect the brain into multiple competing and cooperating generative models, whose synergistic interactions may be productively considered as implementing GAN-type setups. Very speculatively, it may even be the case that perception-action cycles, hemispheric interactions, and interpersonal communication could all be understood as implementing CycleGAN-like dynamics. That is, to what extent could relationships between hemispheres (or between individuals) be analogous to a paired GAN setup, where each system may evaluate the output of the other, so promoting the formation of usefully disentangled representations of features in reduced dimensionality latent spaces [100, 101], thereby promoting controllability and combinatorial power in imagination? These are only some of the many ways that Gershman's intriguing proposal of a "generative adversarial brain" may lead to innovative directions for trying to understand functional relationships within and between minds.

Acknowledgements. I would like to thank the IWAI 2020 organizing committee for arranging this workshop and providing me with an opportunity to share these ideas.

I would like to thank Karl Friston for his mentorship and for being an ongoing source of inspiration over many years.

I would like to thank Amelia Thomley and Victoria Klimaj for their help with illustrations and figures, and for their enduring friendship.

References

1. Friston, K.J., Wiese, W., Hobson, J.A.: Sentience and the origins of consciousness: from cartesian duality to Markovian monism. Entropy **22**, 516 (2020). https://doi.org/10.3390/e22050516
2. Chalmers, D.J.: Facing up to the problem of consciousness. J. Conscious. Stud. **2**, 200–219 (1995)
3. Friston, K.J.: The free-energy principle: a unified brain theory? Nat. Rev. Neurosci. **11**, 127–138 (2010). https://doi.org/10.1038/nrn2787
4. Friston, K.J., FitzGerald, T., Rigoli, F., Schwartenbeck, P., Pezzulo, G.: Active inference: a process theory. Neural Comput. **29**, 1–49 (2017). https://doi.org/10.1162/NECO_a_00912
5. Conant, R.C., Ashby, W.R.: Every good regulator of a system must be a model of that system. Int. J. Syst. Sci. **1**, 89–97 (1970). https://doi.org/10.1080/00207727008920220
6. Hohwy, J., Seth, A.: Predictive processing as a systematic basis for identifying the neural correlates of consciousness. PsyArXiv (2020). https://doi.org/10.31234/osf.io/nd82g
7. Safron, A.: An integrated world modeling theory (IWMT) of consciousness: combining integrated information and global neuronal workspace theories with the free energy principle and active inference framework; toward solving the hard problem and characterizing agentic causation. Front. Artif. Intell. **3** (2020). https://doi.org/10.3389/frai.2020.00030
8. Safron, A.: Integrated world modeling theory (IWMT) revisited. PsyArXiv (2019). https://doi.org/10.31234/osf.io/kjngh
9. Clark, A.: Consciousness as Generative Entanglement. https://www.pdcnet.org/pdc/bvdb.nsf/purchase?openform&fp=jphil&id=jphil_2019_0116_0012_0645_0662. Accessed 13 Jan 2020. https://doi.org/10.5840/jphil20191161241
10. Gazzaniga, M.S.: The Consciousness Instinct: Unraveling the Mystery of How the Brain Makes the Mind. Farrar, Straus and Giroux, New York (2018)
11. Chang, A.Y.C., Biehl, M., Yu, Y., Kanai, R.: Information closure theory of consciousness. arXiv:1909.13045 (2019)
12. Ziporyn, B.: Being and Ambiguity: Philosophical Experiments with Tiantai Buddhism. Open Court, Chicago (2004)
13. Rudrauf, D., Bennequin, D., Granic, I., Landini, G., Friston, K.J., Williford, K.: A mathematical model of embodied consciousness. J. Theor. Biol. **428**, 106–131 (2017). https://doi.org/10.1016/j.jtbi.2017.05.032
14. Williford, K., Bennequin, D., Friston, K., Rudrauf, D.: The projective consciousness model and phenomenal selfhood. Front. Psychol. **9** (2018). https://doi.org/10.3389/fpsyg.2018.02571

15. Metzinger, T.: The Ego Tunnel: The Science of the Mind and the Myth of the Self. Basic Books, New York (2009)
16. Kanai, R., Chang, A., Yu, Y., Magrans de Abril, I., Biehl, M., Guttenberg, N.: Information generation as a functional basis of consciousness. Neurosci. Conscious. **2019** (2019). https://doi.org/10.1093/nc/niz016
17. Corcoran, A.W., Pezzulo, G., Hohwy, J.: From allostatic agents to counterfactual cognisers: active inference, biological regulation, and the origins of cognition (2019). https://doi.org/10.20944/preprints201911.0083.v1
18. Tononi, G., Boly, M., Massimini, M., Koch, C.: Integrated information theory: from consciousness to its physical substrate. Nat. Rev. Neurosci. **17**, 450 (2016). https://doi.org/10.1038/nrn.2016.44
19. Hoel, E.P., Albantakis, L., Marshall, W., Tononi, G.: Can the macro beat the micro? Integrated information across spatiotemporal scales. Neurosci. Conscious. **2016** (2016). https://doi.org/10.1093/nc/niw012
20. Albantakis, L., Marshall, W., Hoel, E., Tononi, G.: What caused what? A quantitative account of actual causation using dynamical causal networks. arXiv:1708.06716 (2017)
21. Klein, B., Hoel, E.: The Emergence of Informative Higher Scales in Complex Networks. https://www.hindawi.com/journals/complexity/2020/8932526/. Accessed 05 April 2020. https://doi.org/10.1155/2020/8932526
22. Hohwy, J.: The self-evidencing brain. Noûs **50**, 259–285 (2016). https://doi.org/10.1111/nous.12062
23. Mashour, G.A., Roelfsema, P., Changeux, J.-P., Dehaene, S.: Conscious processing and the global neuronal workspace hypothesis. Neuron **105**, 776–798 (2020). https://doi.org/10.1016/j.neuron.2020.01.026
24. Whyte, C.J., Smith, R.: The predictive global neuronal workspace: a formal active inference model of visual consciousness. bioRxiv. 2020.02.11.944611 (2020). https://doi.org/10.1101/2020.02.11.944611
25. Safron, A.: The radically embodied conscious cybernetic Bayesian brain: towards explaining the emergence of agency (2019). https://doi.org/10.31234/osf.io/udc42
26. Fries, P.: Rhythms for cognition: communication through coherence. Neuron **88**, 220–235 (2015). https://doi.org/10.1016/j.neuron.2015.09.034
27. Deco, G., Kringelbach, M.L.: Metastability and coherence: extending the communication through coherence hypothesis using a whole-brain computational perspective. Trends Neurosci. **39**, 125–135 (2016). https://doi.org/10.1016/j.tins.2016.01.001
28. Boly, M., Massimini, M., Tsuchiya, N., Postle, B.R., Koch, C., Tononi, G.: Are the neural correlates of consciousness in the front or in the back of the cerebral cortex? Clin. Neuroimaging Evid. J. Neurosci. **37**, 9603–9613 (2017). https://doi.org/10.1523/JNEUROSCI.3218-16.2017
29. Mumford, D.: On the computational architecture of the neocortex. Biol. Cybern. **65**, 135–145 (1991). https://doi.org/10.1007/BF00202389
30. Rao, R.P., Ballard, D.H.: Predictive coding in the visual cortex: a functional interpretation of some extra-classical receptive-field effects. Nat. Neurosci. **2**, 79–87 (1999). https://doi.org/10.1038/4580
31. Bastos, A.M., Usrey, W.M., Adams, R.A., Mangun, G.R., Fries, P., Friston, K.J.: Canonical microcircuits for predictive coding. Neuron **76**, 695–711 (2012). https://doi.org/10.1016/j.neuron.2012.10.038
32. Kingma, D.P., Welling, M.: Auto-encoding variational bayes. arXiv:1312.6114 (2014)
33. Khemakhem, I., Kingma, D.P., Monti, R.P., Hyvärinen, A.: Variational autoencoders and nonlinear ICA: a unifying framework. arXiv:1907.04809 (2020)
34. Marshel, J.H., et al.: Cortical layer–specific critical dynamics triggering perception. Science **365**, eaaw5202 (2019). https://doi.org/10.1126/science.aaw5202

35. Redinbaugh, M.J., et al.: Thalamus modulates consciousness via layer-specific control of cortex. Neuron **106**, 66–75.e12 (2020). https://doi.org/10.1016/j.neuron.2020.01.005

36. Rezaei, H., Aertsen, A., Kumar, A., Valizadeh, A.: Facilitating the propagation of spiking activity in feedforward networks by including feedback. PLoS Comput. Biol. **16**, e1008033 (2020). https://doi.org/10.1371/journal.pcbi.1008033

37. Hesp, C.: Beyond connectionism: a neuronal dance of ephaptic and synaptic interactions: commentary on "The growth of cognition: free energy minimization and the embryogenesis of cortical computation" by Wright and Bourke. Phys. Life Rev. (2020). https://doi.org/10.1016/j.plrev.2020.08.002

38. Northoff, G., Wainio-Theberge, S., Evers, K.: Is temporo-spatial dynamics the "common currency" of brain and mind? In quest of "spatiotemporal neuroscience". Phys. Life Rev. **33**, 34–54 (2020). https://doi.org/10.1016/j.plrev.2019.05.002

39. Berrou, C., Glavieux, A., Thitimajshima, P.: Near Shannon limit error-correcting coding and decoding: Turbo-codes. 1. In: Proceedings of ICC 1993 - IEEE International Conference on Communications, vol. 2, pp. 1064–1070 (1993). https://doi.org/10.1109/ICC.1993.397441

40. McEliece, R.J., MacKay, D.J.C., Cheng, J.-F.: Turbo decoding as an instance of Pearl's "belief propagation" algorithm. IEEE J. Sel. Areas Commun. **16**, 140–152 (1998). https://doi.org/10.1109/49.661103

41. Jiang, Y., Kim, H., Asnani, H., Kannan, S., Oh, S., Viswanath, P.: Turbo autoencoder: deep learning based channel codes for point-to-point communication channels. arXiv:1911.03038 (2019)

42. Parr, T., Markovic, D., Kiebel, S.J., Friston, K.J.: Neuronal message passing using Mean-field, Bethe, and Marginal approximations. Sci. Rep. **9**, 1889 (2019). https://doi.org/10.1038/s41598-018-38246-3

43. Northoff, G., Mushiake, H.: Why context matters? Divisive normalization and canonical microcircuits in psychiatric disorders. Neurosci. Res. (2019). https://doi.org/10.1016/j.neures.2019.10.002

44. Heeger, D.J.: Theory of cortical function. Proc. Natl. Acad. Sci. U.S.A. **114**, 1773–1782 (2017). https://doi.org/10.1073/pnas.1619788114

45. Hawkins, J., Ahmad, S.: Why neurons have thousands of synapses, a theory of sequence memory in neocortex. Front. Neural Circuits **10** (2016). https://doi.org/10.3389/fncir.2016.00023

46. Ahmad, S., Scheinkman, L.: How can we be so dense? The benefits of using highly sparse representations. arXiv preprint. arXiv:1903.11257 (2019)

47. Da Costa, L., Parr, T., Sengupta, B., Friston, K.: Natural selection finds natural gradient. arXiv:2001.08028 (2020)

48. Lillicrap, T.P., Santoro, A., Marris, L., Akerman, C.J., Hinton, G.: Backpropagation and the brain. Nat. Rev. Neurosci. 1–12 (2020). https://doi.org/10.1038/s41583-020-0277-3

49. Vincent, P., Larochelle, H., Lajoie, I., Bengio, Y., Manzagol, P.-A.: Stacked denoising autoencoders: learning useful representations in a deep network with a local denoising criterion. J. Mach. Learn. Res. **11**, 3371–3408 (2010)

50. Hawkins, J., Lewis, M., Klukas, M., Purdy, S., Ahmad, S.: A framework for intelligence and cortical function based on grid cells in the neocortex. Front. Neural Circuits **12** (2019). https://doi.org/10.3389/fncir.2018.00121

51. Kosiorek, A., Sabour, S., Teh, Y.W., Hinton, G.E.: Stacked capsule autoencoders. In: Wallach, H., Larochelle, H., Beygelzimer, A., Alché-Buc, F., Fox, E., Garnett, R. (eds.) Advances in Neural Information Processing Systems, vol. 32. pp. 15512–15522. Curran Associates, Inc., New York (2019)

52. Hu, H.-Y., Li, S.-H., Wang, L., You, Y.-Z.: Machine learning holographic mapping by neural network renormalization group. arXiv:1903.00804 (2019)

53. Li, S.-H., Wang, L.: Neural network renormalization group. Phys. Rev. Lett. **121**, 260601 (2018). https://doi.org/10.1103/PhysRevLett.121.260601

54. Castro, S., El-Deredy, W., Battaglia, D., Orio, P.: Cortical ignition dynamics is tightly linked to the core organisation of the human connectome. PLoS Comput. Biol. **16**, e1007686 (2020). https://doi.org/10.1371/journal.pcbi.1007686

55. Kingma, D.P., Salimans, T., Welling, M.: Variational dropout and the local reparameterization trick. arXiv:1506.02557 (2015)

56. Buzsáki, G., Tingley, D.: Space and time: the hippocampus as a sequence generator. Trends Cogn. Sci. **22**, 853–869 (2018). https://doi.org/10.1016/j.tics.2018.07.006

57. Hawkins, J., Blakeslee, S.: On Intelligence. Times Books, New York (2004)

58. Baldassano, C., Chen, J., Zadbood, A., Pillow, J.W., Hasson, U., Norman, K.A.: Discovering event structure in continuous narrative perception and memory. Neuron **95**, 709–721.e5 (2017). https://doi.org/10.1016/j.neuron.2017.06.041

59. Whittington, J.C., et al.: The Tolman-Eichenbaum machine: unifying space and relational memory through generalisation in the hippocampal formation. bioRxiv. 770495 (2019). https://doi.org/10.1101/770495

60. Moser, E.I., Kropff, E., Moser, M.-B.: Place cells, grid cells, and the brain's spatial representation system. Annu. Rev. Neurosci. **31**, 69–89 (2008). https://doi.org/10.1146/annurev.neuro.31.061307.090723

61. Zhang, F., Li, S., Yuan, S., Sun, E., Zhao, L.: Algorithms analysis of mobile robot SLAM based on Kalman and particle filter. In: 2017 9th International Conference on Modelling, Identification and Control (ICMIC), pp. 1050–1055 (2017). https://doi.org/10.1109/ICMIC.2017.8321612

62. Mannella, F., Gurney, K., Baldassarre, G.: The nucleus accumbens as a nexus between values and goals in goal-directed behavior: a review and a new hypothesis. Front. Behav. Neurosci. **7**, 135 (2013). https://doi.org/10.3389/fnbeh.2013.00135

63. McNamara, C.G., Dupret, D.: Two sources of dopamine for the hippocampus. Trends Neurosci. **40**, 383–384 (2017). https://doi.org/10.1016/j.tins.2017.05.005

64. Koster, R., et al.: Big-loop recurrence within the hippocampal system supports integration of information across episodes. Neuron **99**, 1342–1354.e6 (2018). https://doi.org/10.1016/j.neuron.2018.08.009

65. Hasz, B.M., Redish, A.D.: Spatial encoding in dorsomedial prefrontal cortex and hippocampus is related during deliberation. Hippocampus. n/a. https://doi.org/10.1002/hipo.23250

66. Fraccaro, M., Kamronn, S., Paquet, U., Winther, O.: A disentangled recognition and nonlinear dynamics model for unsupervised learning. In: Advances in Neural Information Processing Systems, pp. 3601–3610 (2017)

67. Hassabis, D., Maguire, E.A.: The construction system of the brain. Philos. Trans. R. Soc. Lond. B Biol. Sci. **364**, 1263–1271 (2009). https://doi.org/10.1098/rstb.2008.0296

68. Faul, L., St. Jacques, P.L., DeRosa, J.T., Parikh, N., De Brigard, F.: Differential contribution of anterior and posterior midline regions during mental simulation of counterfactual and perspective shifts in autobiographical memories. NeuroImage **215**, 116843 (2020). https://doi.org/10.1016/j.neuroimage.2020.116843

69. Canolty, R.T., Knight, R.T.: The functional role of cross-frequency coupling. Trends Cogn. Sci. (Regul. Ed.) **14**, 506–515 (2010). https://doi.org/10.1016/j.tics.2010.09.001

70. Sarel, A., Finkelstein, A., Las, L., Ulanovsky, N.: Vectorial representation of spatial goals in the hippocampus of bats. Science **355**, 176–180 (2017). https://doi.org/10.1126/science.aak9589

71. Hills, T.T.: Neurocognitive free will. Proc. Biol. Sci. **286**, 20190510 (2019). https://doi.org/10.1098/rspb.2019.0510

72. MacKay, D.G.: Remembering: What 50 Years of Research with Famous Amnesia Patient H. M. Can Teach Us about Memory and How It Works. Prometheus Books, Buffalo (2019)

73. Kunz, L., et al.: Hippocampal theta phases organize the reactivation of large-scale electrophysiological representations during goal-directed navigation. Sci. Adv. **5**, eaav8192 (2019). https://doi.org/10.1126/sciadv.aav8192

74. Marr, D.: Vision: A Computational Investigation into the Human Representation and Processing of Visual Information. Henry Holt and Company, New York (1983)

75. Prinz, J.: The intermediate level theory of consciousness. In: The Blackwell Companion to Consciousness, pp. 257–271. Wiley, Hoboken (2017). https://doi.org/10.1002/978111913 2363.ch18

76. Varela, F.J., Thompson, E.T., Rosch, E.: The Embodied Mind: Cognitive Science and Human Experience. The MIT Press, Cambridge (1992)

77. Barsalou, L.W.: Grounded cognition: past, present, and future. Top. Cogn. Sci. **2**, 716–724 (2010). https://doi.org/10.1111/j.1756-8765.2010.01115.x

78. Graziano, M.S.A.: Rethinking Consciousness: A Scientific Theory of Subjective Experience. WWNorton & Company, New York (2019)

79. Zhou, J., et al.: Graph neural networks: a review of methods and applications. arXiv:1812. 08434 (2019)

80. Battaglia, P.W., et al.: Relational inductive biases, deep learning, and graph networks. arXiv: 1806.01261 (2018)

81. Bapst, V., et al.: Unveiling the predictive power of static structure in glassy systems. Nat. Phys. **16**, 448–454 (2020). https://doi.org/10.1038/s41567-020-0842-8

82. Cranmer, M., et al.: Discovering symbolic models from deep learning with inductive biases. arXiv:2006.11287 (2020)

83. Xi, E., Bing, S., Jin, Y.: Capsule network performance on complex data. arXiv preprint arXiv:1712.03480 (2017)

84. Haun, A., Tononi, G.: Why does space feel the way it does? Towards a principled account of spatial experience. Entropy **21**, 1160 (2019). https://doi.org/10.3390/e21121160

85. Haun, A.: What is visible across the visual field? (2020). https://doi.org/10.31234/osf.io/wdpu7

86. Kaplan, R., Friston, K.J.: Planning and navigation as active inference. Biol. Cybern. **112**(4), 323–343 (2018). https://doi.org/10.1007/s00422-018-0753-2

87. Hills, T.T., Todd, P.M., Goldstone, R.L.: The central executive as a search process: priming exploration and exploitation across domains. J. Exp. Psychol. Gen. **139**, 590–609 (2010). https://doi.org/10.1037/a0020666

88. Rochat, P.: Emerging self-concept. In: Bremner, J.G., Wachs, T.D. (eds.) The Wiley-Blackwell Handbook of Infant Development, pp. 320–344. Wiley-Blackwell, Hoboken (2010). https://doi.org/10.1002/9781444327564.ch10

89. Graziano, M.S.A.: The temporoparietal junction and awareness. Neurosci. Conscious. **2018** (2018). https://doi.org/10.1093/nc/niy005

90. Parr, T., Rikhye, R.V., Halassa, M.M., Friston, K.J.: Prefrontal computation as active inference. Cereb. Cortex **30**(2), 395–682 (2019)

91. Stephenson-Jones, M., Samuelsson, E., Ericsson, J., Robertson, B., Grillner, S.: Evolutionary conservation of the basal ganglia as a common vertebrate mechanism for action selection. Curr. Biol. **21**, 1081–1091 (2011). https://doi.org/10.1016/j.cub.2011.05.001

92. Houk, J.C., et al.: Action selection and refinement in subcortical loops through basal ganglia and cerebellum. Philos. Trans. R. Soc. Lond. B Biol. Sci. **362**, 1573–1583 (2007). https://doi.org/10.1098/rstb.2007.2063

93. Humphries, M.D., Prescott, T.J.: The ventral basal ganglia, a selection mechanism at the crossroads of space, strategy, and reward. Prog. Neurobiol. **90**, 385–417 (2010). https://doi.org/10.1016/j.pneurobio.2009.11.003

94. Dabney, W., et al.: A distributional code for value in dopamine-based reinforcement learning. Nature 1–5 (2020). https://doi.org/10.1038/s41586-019-1924-6
95. Morrens, J., Aydin, Ç., Rensburg, A.J. van, Rabell, J.E., Haesler, S.: Cue-evoked dopamine promotes conditioned responding during learning. Neuron **0** (2020). https://doi.org/10.1016/j.neuron.2020.01.012
96. Tani, J.: Exploring Robotic Minds: Actions, Symbols, and Consciousness as Self-Organizing Dynamic Phenomena. Oxford University Press, Oxford (2016)
97. Wang, J.X., et al.: Prefrontal cortex as a meta-reinforcement learning system. Nat. Neurosci. **21**, 860 (2018). https://doi.org/10.1038/s41593-018-0147-8
98. Honkanen, A., Adden, A., Freitas, J. da S., Heinze, S.: The insect central complex and the neural basis of navigational strategies. J. Exp. Biol. 222 (2019). https://doi.org/10.1242/jeb.188854
99. Bengio, Y.: The consciousness prior. arXiv:1709.08568 (2017)
100. Thomas, V., et al.: Independently controllable factors. arXiv preprint arXiv:1708.01289 (2017)
101. Thomas, V., et al.: Disentangling the independently controllable factors of variation by interacting with the world. arXiv preprint arXiv:1802.09484 (2018)
102. Ding, Z., Shao, M., Fu, Y.: Robust multi-view representation: a unified perspective from multi-view learning to domain adaption, pp. 5434–5440 (2018)
103. Friston, K.J., Breakspear, M., Deco, G.: Perception and self-organized instability. Front. Comput. Neurosci. **6** (2012). https://doi.org/10.3389/fncom.2012.00044
104. Paperin, G., Green, D.G., Sadedin, S.: Dual-phase evolution in complex adaptive systems. J. R. Soc. Interface **8**, 609–629 (2011). https://doi.org/10.1098/rsif.2010.0719
105. Humphrey, N.: The Invention of Consciousness. Topoi **39**(1), 13–21 (2017). https://doi.org/10.1007/s11245-017-9498-0
106. Barron, H.C., Auksztulewicz, R., Friston, K.: Prediction and memory: a predictive coding account. Prog. Neurobiol. 101821 (2020). https://doi.org/10.1016/j.pneurobio.2020.101821
107. Pearl, J., Mackenzie, D.: The Book of Why: The New Science of Cause and Effect. Basic Books, New York (2018)
108. Kay, K., et al.: Constant sub-second cycling between representations of possible futures in the hippocampus. Cell. **180**, 552–567.e25 (2020). https://doi.org/10.1016/j.cell.2020.01.014
109. Dohmatob, E., Dumas, G., Bzdok, D.: Dark control: the default mode network as a reinforcement learning agent. Hum. Brain Mapp. **41**, 3318–3341 (2020). https://doi.org/10.1002/hbm.25019
110. Hill, F., Santoro, A., Barrett, D.G.T., Morcos, A.S., Lillicrap, T.: Learning to make analogies by contrasting abstract relational structure. arXiv:1902.00120 (2019)
111. Crouse, M., Nakos, C., Abdelaziz, I., Forbus, K.: Neural analogical matching. arXiv:2004.03573 (2020)
112. Safron, A.: Bayesian analogical cybernetics. arXiv:1911.02362 (2019)
113. Baram, A.B., Muller, T.H., Nili, H., Garvert, M., Behrens, T.E.J.: Entorhinal and ventromedial prefrontal cortices abstract and generalise the structure of reinforcement learning problems. bioRxiv. 827253 (2019). https://doi.org/10.1101/827253
114. Damasio, A.: Self Comes to Mind: Constructing the Conscious Brain. Vintage, New York (2012)
115. Livneh, Y., et al.: Estimation of current and future physiological states in insular cortex. Neuron **0**, (2020). https://doi.org/10.1016/j.neuron.2019.12.027
116. Parascandolo, G., et al.: Divide-and-conquer monte carlo tree search for goal-directed planning. arXiv:2004.11410 (2020)
117. Rueter, A.R., Abram, S.V., MacDonald, A.W., Rustichini, A., DeYoung, C.G.: The goal priority network as a neural substrate of Conscientiousness. Hum. Brain Mapp. **39**, 3574–3585 (2018). https://doi.org/10.1002/hbm.24195

118. Verleger, R., Haake, M., Baur, A., Śmigasiewicz, K.: Time to move again: does the bereitschaftspotential covary with demands on internal timing? Front. Hum. Neurosci. **10** (2016). https://doi.org/10.3389/fnhum.2016.00642

119. Park, H.-D., Barnoud, C., Trang, H., Kannape, O.A., Schaller, K., Blanke, O.: Breathing is coupled with voluntary action and the cortical readiness potential. Nat. Commun. **11**, 1–8 (2020). https://doi.org/10.1038/s41467-019-13967-9

120. Travers, E., Friedemann, M., Haggard, P.: The readiness potential reflects expectation, not uncertainty, in the timing of action. bioRxiv. 2020.04.16.045344 (2020). https://doi.org/10.1101/2020.04.16.045344

121. Adams, R., Shipp, S., Friston, K.J.: Predictions not commands: active inference in the motor system. Brain Struct. Funct. **218**, 611–643 (2013). https://doi.org/10.1007/s00429-012-0475-5

122. Gershman, S.J.: The generative adversarial brain. Front. Artif. Intell. **2** (2019). https://doi.org/10.3389/frai.2019.00018

123. Larsen, A.B.L., Sønderby, S.K., Larochelle, H., Winther, O.: Autoencoding beyond pixels using a learned similarity metric. arXiv:1512.09300 (2016)

124. Ha, D., Schmidhuber, J.: World models. arXiv:1803.10122 (2018). https://doi.org/10.5281/zenodo.1207631

125. Magno, E., Foxe, J.J., Molholm, S., Robertson, I.H., Garavan, H.: The anterior cingulate and error avoidance. J. Neurosci. **26**, 4769–4773 (2006). https://doi.org/10.1523/JNEUROSCI.0369-06.2006

126. Garrison, J.R., Fernyhough, C., McCarthy-Jones, S., Haggard, M., Simons, J.S.: Paracingulate sulcus morphology is associated with hallucinations in the human brain. Nat. Commun. **6**, 8956 (2015). https://doi.org/10.1038/ncomms9956

127. Stolyarova, A., et al.: Contributions of anterior cingulate cortex and basolateral amygdala to decision confidence and learning under uncertainty. Nat. Commun. **10**, 1–14 (2019). https://doi.org/10.1038/s41467-019-12725-1

128. Boroujeni, K.B., Tiesinga, P., Womelsdorf, T.: Interneuron specific gamma synchronization encodes uncertain cues and prediction errors in lateral prefrontal and anterior cingulate cortex. bioRxiv. 2020.07.24.220319 (2020). https://doi.org/10.1101/2020.07.24.220319

129. Lenhart, L., et al.: Cortical reorganization processes in meditation naïve participants induced by 7 weeks focused attention meditation training. Behav. Brain Res. **395**, 112828 (2020). https://doi.org/10.1016/j.bbr.2020.112828

130. Vassena, E., Deraeve, J., Alexander, W.H.: Surprise, value and control in anterior cingulate cortex during speeded decision-making. Nat. Hum. Behav. 1–11 (2020). https://doi.org/10.1038/s41562-019-0801-5

131. Bubb, E.J., Metzler-Baddeley, C., Aggleton, J.P.: The cingulum bundle: anatomy, function, and dysfunction. Neurosci. Biobehav. Rev. **92**, 104–127 (2018). https://doi.org/10.1016/j.neubiorev.2018.05.008

132. Robinson, R.: Stimulating the cingulum relieves anxiety during awake neurosurgery: what is the therapeutic potential? Neurol. Today **19**, 27 (2019). https://doi.org/10.1097/01.NT.0000554700.13747.f2

133. Craig, A.D.B.: Significance of the insula for the evolution of human awareness of feelings from the body. Ann. N. Y. Acad. Sci. **1225**, 72–82 (2011). https://doi.org/10.1111/j.1749-6632.2011.05990.x

134. Schmidhuber, J.: POWERPLAY: training an increasingly general problem solver by continually searching for the simplest still unsolvable problem. arXiv:1112.5309 (2012)

135. Izquierdo-Torres, E., Bührmann, T.: Analysis of a dynamical recurrent neural network evolved for two qualitatively different tasks: walking and chemotaxis. In: ALIFE (2008)

136. Izquierdo, E., Aguilera, M., Beer, R.: Analysis of ultrastability in small dynamical recurrent neural networks. In: The 2018 Conference on Artificial Life: A Hybrid of the European Conference on Artificial Life (ECAL) and the International Conference on the Synthesis and Simulation of Living Systems (ALIFE), vol. 25, pp. 51–58 (2013). https://doi.org/10.1162/978-0-262-31709-2-ch008
137. Pezzulo, G., Rigoli, F., Friston, K.J.: Hierarchical active inference: a theory of motivated control. Trends Cogn. Sci. **22**, 294–306 (2018). https://doi.org/10.1016/j.tics.2018.01.009
138. Richards, B.A., et al.: A deep learning framework for neuroscience. Nat. Neurosci. **22**, 1761–1770 (2019). https://doi.org/10.1038/s41593-019-0520-2
139. Srivastava, N., Hinton, G., Krizhevsky, A., Sutskever, I., Salakhutdinov, R.: Dropout: a simple way to prevent neural networks from overfitting. J. Mach. Learn. Res. **15**, 1929–1958 (2014)
140. Arese Lucini, F., Del Ferraro, G., Sigman, M., Makse, H.A.: How the brain transitions from conscious to subliminal perception. Neuroscience **411**, 280–290 (2019). https://doi.org/10.1016/j.neuroscience.2019.03.047

Confirmatory Evidence that Healthy Individuals Can Adaptively Adjust Prior Expectations and Interoceptive Precision Estimates

Ryan Smith[✉], Rayus Kuplicki, Adam Teed, Valerie Upshaw, and Sahib S. Khalsa

Laureate Institute for Brain Research, Tulsa, OK, USA
rsmith@laureateinstitute.org

Abstract. Theoretical proposals have previously been put forward regarding the computational basis of interoception. Following on this, we recently reported using an active inference approach to 1) quantitatively simulate interoceptive computation, and 2) fit the model to behavior on a cardiac awareness task. In the present work, we attempted to replicate our previous results in an independent group of healthy participants. We provide evidence confirming our previous finding that healthy individuals adaptively adjust prior expectations and interoceptive sensory precision estimates based on task context. This offers further support for the utility of computational approaches to characterizing the dynamics of interoceptive processing.

Keywords: Interoception · Active inference · Precision · Prior expectations · Bayesian perception · Computational modeling

1 Introduction

Multiple neurocomputational models of interoceptive processing have recently been put forward (e.g., [1, 2]). These models have focused largely on understanding interoception within the framework of Bayesian predictive processing models of perception. A central component of such models is the brain's ability to update its model of the body in the face of interoceptive prediction errors (i.e., mismatches between afferent interoceptive signals from the body and prior expectations). To do so adaptively, the brain must also continuously update estimates of both its prior expectations and the reliability (precision) of afferent sensory signals arising from the body. In a recent study [3], we described a formal generative model based on the active inference framework that simulated approximate Bayesian perception within a cardiac perception (heartbeat tapping) task. We fit this model to behavioral data and found evidence that healthy individuals successfully adapted their prior expectations and sensory precision estimates during different task contexts, particularly under conditions of interoceptive perturbation. In contrast, a transdiagnostic psychiatric sample showed a more rigid pattern in which precision estimates remained stable across task conditions. As this study was the first to present such evidence, confirmatory evidence is lacking. In the present study, we attempted to replicate

© Springer Nature Switzerland AG 2020
T. Verbelen et al. (Eds.): IWAI 2020, CCIS 1326, pp. 156–164, 2020.
https://doi.org/10.1007/978-3-030-64919-7_16

the previous finding in healthy participants by fitting our model to behavior on the same task in a new sample. As in our previous study, we assessed cardiac interoceptive awareness under resting conditions with different instruction sets where 1) guessing was allowed, and 2) guessing wasn't allowed; we also 3) assessed performance during an interoceptive perturbation (inspiratory breath-hold) condition expected to increase the precision of the afferent cardiac signal (while also under the no-guessing instruction). We predicted that prior expectations for feeling a heartbeat would be reduced under the no-guessing instruction and that sensory precision estimates would increase in the breath-hold condition relative to the resting conditions. We also sought to confirm continuous relationships we previously observed between these model parameters and two facets of interoceptive awareness: self-reported heartbeat intensity (positive relationship with both parameters) and self-reported task difficulty (negative relationship with both parameters).

2 Methods

Data were collected from a community sample of 63 participants (47 female; mean age = 24.94, SD = 6.09) recruited via advertisements and from an existing database of participants in previous studies. Participants were screened using the Mini International Neuropsychiatric Inventory 6 or 7 (MINI) and did not meet criteria for any disorder. Our initial assessment identified some participants with poor electrocardiogram (EKG) traces, which were removed from our analyses. Final sample sizes for each condition are shown in Table 1.

Participants completed the same cardiac perception ("heartbeat tapping") task as in our previous study [3], wherein participants were asked to close their eyes and press down on a key each time they felt their heartbeat, and to try to mirror their heartbeat as closely as possible. Participants were not permitted to take their pulse (e.g., hold their finger to their wrist or neck) or to hold their hand against their chest. Thus, they could only base their choice to tap on their internally felt sensations. The task was repeated under multiple conditions designed to assess the influence of cognitive context and physiological perturbation on performance. In the first condition, participants were told that, even if they weren't sure about what they felt, they should take their best guess ("guessing condition"). This condition was included because it matches a standard instruction given during heartbeat counting tasks [4]. In the second condition, they were told to only press the key when they actually felt their heartbeat, and if they did not feel their heartbeat then they should not press the key (the "no-guessing" condition). In other words, unlike the first time they completed the task, they were specifically instructed not to guess if they didn't feel anything. This condition can be seen as placing an additional cognitive demand on the participant to monitor their own confidence in whether a heartbeat was actually felt; such instructions have been reported to substantially influence performance on the heartbeat counting task [5, 6]. Finally, in the perturbation condition, participants were again instructed not to guess but were also asked to first empty their lungs of all air and then inhale as deeply as possible and hold it for as long as they could tolerate (up to the length of the one-minute trial) while reporting their perceived heartbeat sensations. This third condition (the "breath-hold" condition) was used in an attempt to increase the

Table 1. Mean and standard deviation of study variables by task condition.

	Guessing	No-guessing	Breath-hold	Tone	p-value*
n	50	50	49	50	
Demographic variables					
Age	24.42 (5.93)	24.42 (5.93)	24.55 (5.91)	24.42 (5.93)	ns
Gender (Male)	13 (26.0%)	13 (26.0%)	13 (26.5%)	13 (26.0%)	ns
BMI	24.15 (3.31)	24.15 (3.31)	24.13 (3.35)	24.15 (3.31)	ns
Task variables					
Heart rate	70.94 (9.9)	69.76 (9.43)	70.06 (9.62)	71.1 (9.74)	ns
Taps	51.26 (19.90)	16.12 (19.18)	27.14 (21.32)	77.60 (1.69)	<0.001****
IP	0.04 (0.03)	0.05 (0.05)	0.07 (0.07)	0.17 (0.12)	0.005***
pHB	0.32 (0.12)	0.12 (0.10)	0.18 (0.12)	0.50 (0.01)	<0.001****
Difficulty	51.42 (27.60)	50.88 (33.83)	48.43 (28.89)	20.06 (19.02)	ns
Confidence	26.24 (19.96)	44.42 (31.51)	52.49 (27.37)	74.50 (16.59)	<0.001**
Intensity	19.70 (17.89)	15.54 (15.95)	42.12 (28.30)	83.26 (16.48)	<0.001***
Counting accuracy	0.67 (0.23)	0.23 (0.27)	0.39 (0.31)	0.99 (0.02)	<0.001****

*These p-values are based on linear mixed effects analyses (LMEs) that exclude the Tone condition. For task variables (except heart rate), analyses also included age, gender, BMI, precision estimates in the tone condition, heart rate, and its interaction with task condition as covariates.
**Guessing condition significantly differed from the other heartbeat tapping conditions.
***Breath-hold condition significantly differed from the other heartbeat tapping conditions.
****All heartbeat tapping conditions were significantly different from one another.

strength of the afferent cardiac signal by increasing physiological arousal. We expected 1) that cardiac perception would be poor in the guessing condition (i.e., as only roughly 35% of individuals appear to accurately perceive their heartbeats under resting conditions [7]), 2) that tapping would be more conservative in the no-guessing condition, and 3) that the breath-hold condition would result in improved performance on average (i.e., as interoceptive accuracy has been shown to increase under conditions of heightened cardiorespiratory arousal [8–10]). Directly after completing each task condition, participants were asked to rate subjective task difficulty, performance, and heartbeat intensity from 0 to 100. Participants also completed a control condition in which they tapped every time they heard a 1000 Hz auditory tone presented for 100 ms (78 tones, randomly jittered by ±10% and presented in a pattern following a sine curve with a frequency of 13 cycles/minute, mimicking the range of respiratory sinus arrhythmia during a normal breathing rate of 13 breaths per minute). This was completed between the first (guessing) and second (no-guessing) heartbeat tapping conditions. As body mass index (BMI) is a potential confound, we also measured this for each participant.

A three-lead EKG was used to assess the objective timing of participants' heartbeats throughout the task. The pulse oximeter signal was also gathered using a pulse plethysmography (PPG) device attached to the ear lobe. These signals were acquired simultaneously on a Biopac MP150 device. Response times were collected using a task implemented in PsychoPy, with data collection synchronized via a parallel port interface. EKG and response data were scored using in-house developed MATLAB code.

To model behavior, we divided each task time series into intervals corresponding to windows equally dividing the time period between each heartbeat, based on each participant's EKG recording. Potentially perceivable heartbeats were specifically based on the timing of the peak of the EKG R-wave (signaling electrical depolarization of the atrioventricular neurons of the heart) +200 milliseconds (ms). This 200 ms interval was considered a reasonable estimate of participants' pulse transit time (PTT) according to previous estimates for the ear PTT [11], which signals the mechanical transmission of the systolic pressure wave to the earlobe – and was considered a lower bound on how quickly a heartbeat could be felt (and behaviorally indicated) after it occurred. We also confirmed this by measuring the PTT of each participant, defined as the distance between the peak of the EKG R-wave and the onset of the peak of the PPG waveform (usable quality median PTT values were available in 45 participants; mean $= 200$ ms, SD $= 2$ ms). The length of each heartbeat interval (i.e., the "before-beat interval" and "after-beat interval") depended on the heart rate. For example, if two heartbeats were 1 s apart, the "after-beat interval" would include the first 500 ms after the initial beat and the "before-beat interval" would correspond to the 2nd 500 ms. The after-beat intervals were considered the time periods in which the systole (heart muscle contraction) signal was present and in which a tap should be chosen if it was felt. The before-beat intervals were treated as the time periods where the diastole (heart muscle relaxation) signal was present and in which tapping should not occur (i.e., assuming taps are chosen in response to detecting a systole; e.g., as supported by [12]). This allowed us to formulate each interval as a "trial" in which either a tap or no tap could be chosen and in which a systole or diastole signal was present (see Fig. 1). Each trial formally consisted of two time points. At the first time point, the model always began in a "start" state with an uninformative "start" observation. At the second time point, either a systole or diastole observation was presented, based on whether that trial corresponded to the time window before or after a systole within the participant's EKG signal (as described above). The model then inferred the probability of the presence vs. absence of a heartbeat (corresponding to the probability of choosing whether or not to tap). At this point, the trial ended, and the next trial began with the model again beginning in the "start" state and being presented with a new systole or diastole signal, and so forth. For further details on all methods, see [3].

To model behavior, we used a Bayesian generative model of perception (see Fig. 1) derived from the Markov decision process (MDP) formulation of active inference [13]. Unlike the full MDP model, however, we only explicitly included a generative model of perception. Observations (o) included systole, diastole, and a "start" observation (i.e., based on each individual's EKG recording). These observations were generated by hidden (perceptual) states (s) that included either feeling one's heartbeat or not, as well as a "start" state. The probability of choosing to tap on each trial was assumed to correspond to the posterior probability of the heartbeat state on each trial. Here, a trial formally included two timesteps: 1) a "start" time point, followed by 2) the possibility of either a systole or diastole. The matrices and equations defining the model are specified in Fig. 1. This model was used in conjunction with the standard SPM_MDP_VB_X routine (within the freely available SPM12 software package; Wellcome Trust Centre

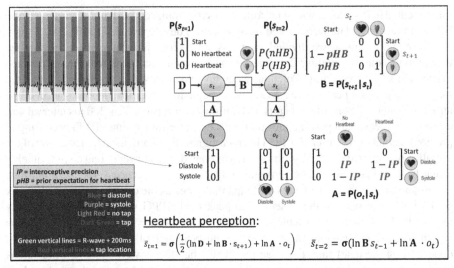

Fig. 1. (**Upper Right**) A graphical depiction of the computational model. This is a simplified version of a commonly used active inference formulation of partially observable Markov decision processes [13], which does not explicitly model action. Systole/diastole signals (derived from EKG; **Upper Left**) were modeled as observations, and beliefs about the presence or absence of a heartbeat were modeled as hidden states. For simplicity, model-fitting assumed that the probability of choosing to tap corresponded to the posterior distribution over states ($\bar{s}$) – that is, the relative confidence in the presence vs. absence of a heartbeat: P(*HB*) and P(*nHB*), respectively. Estimated model parameters included: 1) interoceptive precision (*IP*) – the precision of the mapping from systole/diastole to beliefs about heartbeat/no heartbeat in the **A** matrix, which can be associated with the weight assigned to sensory prediction errors; and 2) prior expectations for the presence of a heartbeat (*pHB*). Because minimal precision corresponds to an *IP* value of 0.5, and both higher and lower values indicate that taps more reliably track systoles (albeit in an anticipatory or reactive manner), our ultimate measure of precision subtracted 0.5 from raw *IP* values and then took their absolute value. On each trial, beliefs about the probability of a heartbeat (corresponding to the probability of choosing to tap) relied on Bayesian inference as implemented in the "heartbeat perception" equations (**Bottom Right**). Note that, by convention in active inference, the dot product (·) applied to matrices here indicates transposed matrix multiplication, and σ denotes a softmax (normalized exponential) function (see text for details).

for Neuroimaging, London, UK, http://www.fil.ion.ucl.ac.uk/spm) to simulate cardiac response data.

The precision of the likelihood matrix **A** was controlled by an "interoceptive precision" (*IP*) parameter. Prior expectations were controlled by a parameter *pHB* within the transition matrix **B**. Note that, because each "trial" was based on equally dividing the time periods before and after each heartbeat (i.e., based on each individual's EKG; resulting in alternating "systole" and "diastole" trials), this entails that the "correct" *pHB* value would be 0.5. Both *IP* and *pHB* were estimated for each participant by finding values that maximized the likelihood of their responses using variational Laplace – that is, values that maximized the posterior probability of the heartbeat state on trials in which they chose to tap (implemented by the spm_nlsi_Newton.m parameter estimation routine available within SPM). Prior means and variances for each parameter were both set to 0.5. Because "raw" IP values (IP_{raw}) both above and below 0.5 indicate higher precision (i.e., values approaching 0 indicate reliable anticipatory tapping, whereas values approaching 1 indicate reliable tapping after a systole), our ultimate measure of precision was recalculated by centering IP_{raw} on 0 and taking its absolute value.

Primary confirmatory analyses included linear mixed effects analyses (LMEs) assessing the main effect of task condition on each parameter, while accounting for age, gender, BMI, heart rate, and its interaction with task condition. To help rule out the possibility that IP estimates were driven by differences in motor stochasticity, we also included precision estimates for the tone condition as an additional covariate. This was based on the assumption that, because the sensory signal in the tone condition is highly precise, any variability in precision estimates in the tone condition would be better explained by individual differences in random influences on behavior as opposed to perception.

3   Results

As in our previous study, an LME (excluding the tone condition) revealed a main effect of task condition on IP ($F(2,95) = 5.65$, $p = .005$), after accounting for age, gender, BMI, precision in the tone condition, heart rate, and its interaction with task condition (Fig. 2). Post-hoc Tukey comparisons indicated that IP was significantly greater in the breath-hold condition than in the guessing ($p = .006$) and no-guessing ($p = .028$) conditions. An identical analysis focused on pHB revealed the expected effect of task condition ($F(2,95) = 56.18$, $p < .001$), in which 1) pHB was significantly lower in the no-guessing and breath-hold conditions than in the guessing condition ($ps < .001$; note that the breath-hold condition still included the no-guessing instruction), and 2) it was higher in the breath-hold condition than in the no-guessing condition ($p = .01$).

Secondary analyses examined the relationships between model parameters and other task variables at a threshold of $p < .01$, uncorrected (shown in Fig. 3). These results largely confirmed the relationships observed in our earlier study [3], including positive relationships between both IP and pHB parameters and self-reported heartbeat intensity, and negative relationships between these parameters and self-reported task difficulty. Note that expected relationships with difficulty in the breath-hold condition were not significant in this sample at our stated threshold of $p < .01$, but were significant at a more liberal threshold of $p < .05$ and had very similar correlation magnitudes as in our previous results, which were significant in that larger sample. We also confirmed that model parameters were not correlated with individual differences in median PTT.

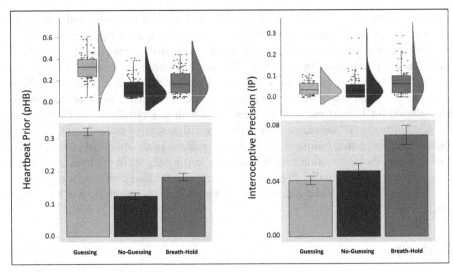

Fig. 2. Bottom: Mean and standard error for prior expectations (*pHB*; **Left**) and interoceptive precision (*IP*) estimates (**Right**) by condition. **Top:** Raincloud plots depicting the same results in terms of individual datapoints, boxplots (median and upper/lower quartiles), and distributions. *pHB* was significantly lower in the no-guessing and breath-hold conditions than in the guessing condition (*p*s < .001) and it was higher in the breath-hold condition than in the no-guessing condition (*p* = .01). *IP* was significantly greater in the breath-hold condition than in the guessing (*p* = .006) and no-guessing (*p* = .028) conditions.

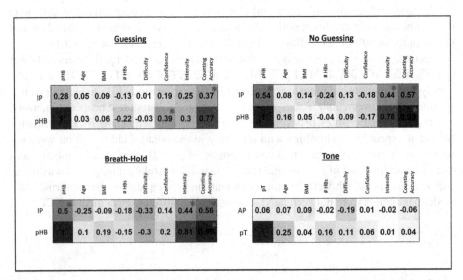

Fig. 3. Pearson correlations between model parameters and self-report and other task-relevant variables for each task condition across all participants. *IP* = interoceptive precision parameter, *pHB* = prior expectation for heartbeat parameter, *pT* = prior expectation for tone parameter, *AP* = auditory precision, #HBs = number of heartbeats during the task condition, BMI = body mass index. For reference, correlations at *p* < .01 (uncorrected) are marked with red asterisks.

4   Discussion

In this study we sought, and found, confirmatory evidence that an interoceptive (inspiratory breath-hold) perturbation increased the precision estimates assigned to cardiac signals in healthy individuals. The effectiveness of the perturbation was further validated by the finding that participants reported more intense heartbeat sensations in the breath-hold condition (see Table 1). We further confirmed that prior expectations to feel a heartbeat were reduced when individuals were given a no-guessing instruction, and that both parameters correlated with self-report measures in predicted directions. This replication represents an important step towards empirically advancing our understanding of the computational dynamics underlying interoception. Future work remains to confirm our other previous finding – that interoceptive precision is *not* adjusted across conditions within psychiatric disorders [3]. If this latter result is replicated in future work, it would support the use of our novel interoceptive modelling and model-fitting approach as an important new avenue for computationally phenotyping patient populations at the individual level.

References

1. Barrett, L., Simmons, W.: Interoceptive predictions in the brain. Nat. Rev. Neurosci. **16**, 419–429 (2015)
2. Smith, R., Thayer, J.F., Khalsa, S.S., Lane, R.D.: The hierarchical basis of neurovisceral integration. Neurosci. Biobehav. Rev. **75**, 274–296 (2017)
3. Smith, R., Kuplicki, R., Feinstein, J., Forthman, K.L., Stewart, J.L., Paulus, M.P., et al.: A Bayesian computational model reveals a failure to adapt interoceptive precision estimates across depression, anxiety, eating, and substance use disorders. PLoS Comput. Biol. (2020, in press)
4. Pollatos, O., Herbert, B.M., Matthias, E., Schandry, R.: Heart rate response after emotional picture presentation is modulated by interoceptive awareness. Int. J. Psychophysiol. **63**(1), 117–124 (2007)
5. Desmedt, O., Luminet, O., Corneille, O.: The heartbeat counting task largely involves non-interoceptive processes: evidence from both the original and an adapted counting task. Biol. Psychol. **138**, 185–188 (2018)
6. DeVille, D.C., Kuplicki, R., Stewart, J.L., Tulsa, I., Aupperle, R.L., Bodurka, J., et al.: Diminished responses to bodily threat and blunted interoception in suicide attempters. Elife **9**, e51593 (2020)
7. Khalsa, S.S., Lapidus, R.C.: Can interoception improve the pragmatic search for biomarkers in psychiatry? Front. Psychiatry **7**, 121 (2016)
8. Khalsa, S.S., Rudrauf, D., Sandesara, C., Olshansky, B., Tranel, D.: Bolus isoproterenol infusions provide a reliable method for assessing interoceptive awareness. Int. J. Psychophysiol. **72**(1), 34–45 (2009)
9. Hassanpour, M.S., Yan, L., Wang, D.J., Lapidus, R.C., Arevian, A.C., Simmons, W.K., et al.: How the heart speaks to the brain: neural activity during cardiorespiratory interoceptive stimulation. Philos. Trans. R. Soc. Lond. B Biol. Sci. **371**(1708), 20160017 (2016)
10. Schandry, R., Bestler, M., Montoya, P.: On the relation between cardiodynamics and heartbeat perception. Psychophysiology **30**(5), 467–474 (1993)
11. Allen, J., Murray, A.: Age-related changes in the characteristics of the photoplethysmographic pulse shape at various body sites. Physiol. Meas. **24**(2), 297–307 (2003)

12. Ring, C., Brener, J.: The temporal locations of heartbeat sensations. Psychophysiology **29**(5), 535–545 (1992)
13. Da Costa, L., Parr, T., Sajid, N., Veselic, S., Neacsu, V., Friston, K.: Active inference on discrete state-spaces – a synthesis. arXiv:2001.07203v2 (2020)

Visual Search as Active Inference

Emmanuel Daucé[1,2]([⊠]) [iD] and Laurent Perrinet[1] [iD]

[1] Institut de Neurosciences de la Timone, CNRS/Aix-Marseille Univ,
Marseille, France
emmanuel.dauce@univ-amu.fr
[2] Ecole Centrale Marseille, Marseille, France

Abstract. Visual search is an essential cognitive ability, offering a prototypical control problem to be addressed with Active Inference. Under a Naive Bayes assumption, the maximization of the information gain objective is consistent with the separation of the visual sensory flow in two independent pathways, namely the "What" and the "Where" pathways. On the "What" side, the processing of the central part of the visual field (the fovea) provides the current interpretation of the scene, here the category of the target. On the "Where" side, the processing of the full visual field (at lower resolution) is expected to provide hints about future central foveal processing given the potential realization of saccadic movements. A map of the classification accuracies, as obtained by such counterfactual saccades, defines a utility function on the motor space, whose maximal argument prescribes the next saccade. The comparison of the foveal and the peripheral predictions finally forms an estimate of the future information gain, providing a simple and resource-efficient way to implement information gain seeking policies in active vision. This dual-pathway information processing framework is found efficient on a synthetic visual search task with a variable (eccentricity-dependent) precision. More importantly, it is expected to draw connections toward a more general actor-critic principle in action selection, with the accuracy of the central processing taking the role of a value (or intrinsic reward) of the previous saccade.

Keywords: Object detection · Active Inference · Visual search ·
Visuomotor control · Deep learning

1 Introduction

Moving fast the eye toward relevant regions of the scene interestingly combines elements of action selection (moving the eye) with visual information processing. Noteworthy, the visual sensors have evolved during natural selection toward maximizing their efficiency under strong energy constraints. Vision in most mammals, for instance, has evolved toward a foveated sensor, maintaining a high density of photoreceptors at the center of the visual field, and a much lower density at the periphery. This limited bandwidth transmission is combined with a high mobility of the eye, that allows to displace the center of sight toward different

© Springer Nature Switzerland AG 2020
T. Verbelen et al. (Eds.): IWAI 2020, CCIS 1326, pp. 165–178, 2020.
https://doi.org/10.1007/978-3-030-64919-7_17

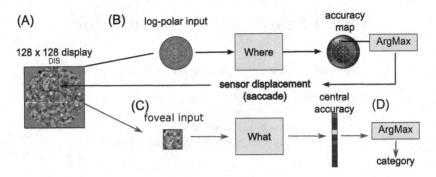

Fig. 1. Computational graph. Based on the anatomy of mammals' visual pathways, we define the following stream of information to implement visual search, one stream for localizing the object in visual space ("Where?"), the other for identifying it ("What?"). **(A)** The visual display is a stack of three layers: first a natural-like background noise is generated, characterized by noise contrast, mean spatial frequency and bandwidth [1]. Then, a sample digit is selected from the MNIST dataset [2], rectified, multiplied by a contrast factor and overlaid at a random position. Last, a circular, gray mask is put on. **(B)** The visual display is then transformed in a retinal input which is fed to the "Where" pathway. This observation is generated by a bank of filters whose centers are positioned on a log-polar grid and whose size increases proportionally with the eccentricity. The "Where" network outputs a collicular-like accuracy map. It is implemented by a three-layered neural network consisting of the retinal log-polar input, two hidden layers (fully-connected linear layers combined with a ReLU non-linearity) with 1000 units each. This map has a similar log-polar (retinotopic) organization and predicts the accuracy at the counter-factual positions of affordable saccades. The position of maximal activity in the "Where" pathway serves to generate a saccade denoted which displaces the center of gaze at a new position. **(C)** This generates a new sensory input in the fovea which is fed to a classification network ("What" pathway). This network is implemented using the three-layered LENET neural network [2]. This network outputs a vector predicting the accuracy of detecting the correct digit. **(D)** Depending on the (binary) success of this categorical identification, we can first reinforce the What network, by supervisedly learning to associate the output with the ground truth through back-propagation. Then, we similarly train the "Where" network by updating its approximate prediction of the accuracy map.

parts of the visual scene, at up to 900 degrees per second in humans. Beyond the energetic efficiency, foveated vision improves the performance of agents by allowing them to focus on relevant vs. irrelevant information [3]. As such, this action perception loop uniquely specifies an AI problem [4,5].

Indeed, Friston [6] proposed the FEP as a general explanatory principle behind the puzzling diversity of the mechanistic processes taking place in the brain and the body. One key ingredient to this process is the (internal) representation of counterfactual predictions, that is, the probable consequences of possible hypothesis as they would be realized into actions (here, saccades). Equipping the agent with the ability to actively sample the visual world allows to interpret saccades as optimal experiments, by which the agent seeks to confirm predictive models of the (hidden) world [4,7]) Following such an active inference scheme, numerical simulations reproduce sequences of eye movements that fit well with empirical data [8,9].

In particular, we focus here on *visual search* which is the cognitive ability to locate a single visual object in cluttered visual scene by placing the fovea on the object, in order to identify it [10–12]. As such, visual search intimately links the sampling of visual space (as it is done by the sensory apparatus) to the behavior which directs this sampling through the action of moving the direction of gaze. Note that the retina samples visual information predominantly on the fovea, though the target may lie in the periphery, where the acuity is lower. It is therefore commonplace that the target is not identifiable with the current information contained on the retinal image. As a consequence, visual search involves the problem that, given a limited observability, the object has to be localized *before* being identified.

Compared to earlier modelling studies, such as [13], we are concerned with the problem of both locating *and* identifying the target. This implies the capability to process the visual data and extract features from a complex (non-uniform) retinotopic visual sampling. This observation highlights an important hypothesis for solving the visual search problem. The semantic content of a visual scene is indeed defined by the positions and identities of the many objects that it contains. In all generality, the identity of an object is independent from its position in retinotopic space which is contingent on the observer's point of view. We thus consider the assumption that the visual system of mammals is built around such an independence hypothesis. The independence assumption, largely exploited in machine learning, is also known as the "Naïve Bayes" assumption. It simply considers as independent the different factors (or latent features) that explain the data. This implies here that inferring the identity and the position can be performed independently, and thus, could be processed *sequentially*. Selecting an object and identifying both its position and category may thus be the elementary bricks of visual processing. It may moreover explain the general separation of visual processing into the ventral and dorsal pathways. These two specific processing pathways are devoted to the processing of the stream of visual information, either to identify the semantic content of the visual field (the "What" pathway), or to decide where to orient next the line of sight (the "Where" pathway). They may operate in a continual and incremental turn-taking fashion, contributing to understand and exploit at best the visual information.

2   Problem Statement: Formalizing Visual Search as Accuracy Seeking

2.1   Visual Search Task

In this manuscript, we built upon an existing model [14] by precisely defining the mathematical framework under the Active Inference formalism. This model is based on a simplified generative model for a visual search task and a proposed algorithm to implement the task. First, in order to implement those principles into a concrete image processing task, we construct a simple yet ecological virtual experiment: After a fixation period of 200 ms, an observer is presented with a luminous 128×128 display showing a single target overlaid on a realistic noisy background (see Fig. 1A). This target is drawn in our case from the MNIST database of manuscript digits consisting of 60000 grayscale images of size 28×28 [2]. This image is displayed for a short period of about 500 ms which allows to perform (at most) one saccade toward the (unique) target. The goal of the agent is ultimately to correctly identify the digit.

2.2   Central Processing

Following the Free Energy minimization principle (FEP) [6], engaging in a saccade stems on maintaining the visual field within the least surprising possible state. This implies, for instance, the capability to predict the next visual input through a generative model, and to orient the sight toward regions that minimize the agent's predicted model surprise [4]. Due to their limited memory and processing capabilities, living brains do not afford to predict or simulate their sensory environment exhaustively. Given the vast diversity of possible visual fields, one should assume that only the foveated part should deserve predictive coding. This implies that the saccadic motor control should be tightly optimized in order to provide a foveal data that should allow to accurately identify (and predict) the target.

   In our model, we divide the retina into the fovea, which constitutes the center of the retina, and the peripheral region, which provides a visual information with a decreasing precision as a function of eccentricity. When considering the full visual field, the exponential decrease of the density of photo-receptors with respect to eccentricity [15] must be reflected in a non-uniform sampling of the visual data. It is here implemented as a log-polar conformal mapping, as it provides a good fit with observations in mammals and has a long history in computer vision and robotics [16]. These coordinates are denoted as the couple $u = (\epsilon, \theta)$ corresponding respectively to the log-eccentricity and azimuth in (spherical) polar coordinate by $\rho(u) \stackrel{\text{def.}}{=} (R \cdot \exp(\epsilon) \cdot \cos \theta, R \cdot \exp(\epsilon) \cdot \sin \theta)$ with R the maximal eccentricity.

   Let us define as x a spatial coordinate in the input image cartesian referential (with $x_0 \stackrel{\text{def.}}{=} (0,0)$ defining the center of the image), with x^t the position of the target and $k^t \in \{0, \ldots, 9\}$ its identity. The content of the fovea is considered as spatially uniform, here defined by extracting the 28×28 sub-image $f^t(x)$ at gaze direction x (initially x_0). At any given trial t drawn from the set $\mathcal{T}$ of trials of our virtual experiment, knowing the corresponding position x^t of the object, the problem of identifying the object can be solved, for instance, by a deep neural network [2] which infers its category. This network takes $f^t(x)$ as an input and returns a multinomial distribution vector $\mathbf{a}(f^t(x)) \in \mathbb{R}^{10}$ (with $\sum_k \mathbf{a}_k(f^t(x)) = 1$). This network takes here the role of the "What" pathway. Knowing the correct label k^t (and position x^t) for this trial t, this network is trained using a gradient descent with a categorical Cross Entropy loss. This loss is by definition:

$$\mathcal{L}_{\mathcal{K}}^t = -\log \mathbf{a}_{k^t}(f^t(x)) \tag{1}$$

The gradient descent is computed at each trial and the process is iterated over the set $\mathcal{T}$ of trials which give pairs of inputs $f^t(x)$ and outputs k^t for this supervised learning scheme. The accuracy of this classic neural network is known to exceed 98% over the genuine MNIST database [2], on par with human performance. Due to the max-pooling layers used between the convolutional layers, it also shows a robust translation invariance. In our experimental conditions, the network is trained over an augmented MNIST digits dataset, having a variable contrast, a variable shift (from 0 to 15 pixels away from the center) and a variable (randomly generated) background.

Knowing $f^t(x)$, the categorical response is $k^t = \arg\max_k \mathbf{a}_k(f^t(x))$. This response can be correct or incorrect. The correctness of the response is noted $\mathbf{o}(f^t(x))$ as we test our model. This value, that is 1 for a correct response and 0 otherwise, can be interpreted as a binary random variable. This random variable can be sampled at different t, with different success or failures depending on the actual target position x^t.

2.3   Accuracy Map

The "What" neural network is constructed such that it can provide an estimate of the chance of success for every possible category by processing the central part of the visual field, i.e. the fovea. This chance of success could in principle be estimated the same way at any peripheral position $x \neq x_0$, through making a saccade and estimating the chance of success at gaze direction x. Then, for any target position x^t, and under an ergodic assumption, it could provide a belief on the average success that would be obtained at all positions x, i.e. $A^t(x) \overset{\text{def.}}{=} \mathbf{a}_{k^t}(f^t(x)) \approx Pr(k^t|f^t(x))$. This accuracy being defined for any gaze direction x, one could thus construct a *map* providing the expected probability of classification success knowing a potential future eye direction x afforded by a saccade. From the definition of the "What" network, this could be simply approximated by the accuracy of the selected class:

$$A^t(x) \approx \max_k \mathbf{a}_k(f^t(x)) \tag{2}$$

In principle, one could extract all possible sub-images $f^t(x)$ at all positions x, and estimate $A^t(x)$ directly. Moving the eye toward $\hat{x} = \arg\max_x A^t(x)$ and finding the object's identity at location $\hat{x}$ would solve the problem of both identifying and locating the target. This brute-force solution is of course computationally prohibitive, but provides a baseline toward a more biologically-relevant processing.

The belief in the success or the failure of identifying the target at different positions being, by construction, an output of the "What" pathway, it is essential for a visual search task to estimate the correctness of the test *prior* to a saccade, that is, to predict the statistics of $\mathbf{o}(f^t(x))$ from $A^t(x)$. The eye next position x being the result of a motor displacement u, with $x = x_0 + \rho(u) = \rho(u)$, it should be governed by a *policy*, i.e. a method that selects the next movement from the available visual input. The set of all possible displacements forms a *motor map*, and such a policy can be formalized as a mapping from the visual input space toward the motor map. Following the classical reinforcement learning literature, the motor map is expected to provide a value over the space of actions. We postulate here that *the value of the motor displacement u is identified with the classification accuracy obtained at position $x = \rho(u)$.* Moreover, we will show that, with minimal simplifying assumptions, this postulate can be framed into the more general framework of Active Inference.

3   Principles: Supervised Learning of Action Selection

3.1   Peripheral Visual Processing

On the visual side, local visual features are extracted as oriented edges as a combination of the retinotopic transform with filters resembling that found in the primary visual cortex [17]. The centers of these filters are radially organized around the center of fixation, with small receptive fields at the center and more large and scarce receptive fields at the periphery. The size of the filters increases proportionally with the eccentricity. To cover the visual space from the periphery to the fovea, we used 10 spatial eccentricity scales $\epsilon \in [-4, -1]$ such that the filters are placed at about 2, 3, 4.5, 6.5, 9, 13, 18, 26, 36.5, and 51.3 pixels from the center of gaze. There are 24 different azimuth angles allowing them to cover most of the original 128×128 image. At each of these positions, 6 different edge orientations and 2 different phases (symmetric and anti-symmetric) are computed.

This finally implements a bank of linear filters which models the receptive fields of the primary visual cortex. Assuming this log-polar arrangement, the resulting retinal visual data at this trial is noted as the feature vector $\mathbf{s}^t(x)$. For simplicity, it is noted $\mathbf{s}^t$ further on. The length of this vector is 2880, such that this retinal processing compresses the original image by about 83%, with high spatial frequencies preserved at the center and only low spatial frequencies conserved at the periphery. In practice, the bank of filters is pre-computed and placed into a matrix for a rapid transformation of input batches into feature vectors.

3.2   Motor Control

Assuming the motor control is independent from the identity pathway, we take the classification success, as measured at the output of the "What" pathway, as the principal outcome of the "Where" pathway. It is assumed, in short, that the surprise should be higher in case of failure than in case of success, and that minimizing the surprise through active inference should be consistent with maximizing the likelihood of success.

On the motor side, a possible saccade location is defined as $u \stackrel{\text{def.}}{=} (\epsilon, \theta)$. Each coordinate of the visual field, except for the center, is mapped on a saccadic motor map. The motor map is also organized radially in a log-polar fashion, making the control more precise at the center and coarser at the periphery. This modeling choice is reminiscent of the approximate log-polar organization of the superior colliculus (SC) motor map [18]. Given a saccade command u, the corresponding classification success is noted $\mathbf{o}(f^t(\rho(u)))$. This success (or failure) being measured after the saccade, it must be guessed from a model. We posit here that the principle underlying the "Where" processing pathway is to predict the probability of success for every possible saccade command. This success is considered a realization of the likelihood $p(\mathbf{o}|u, \mathbf{s}^t)$. It is important here to note the dependence on the (peripheral) visual observation $\mathbf{s}^t$. Our likelihood function p can be seen as a mapping from $\mathbf{s}^t$ to the set $\mathcal{U}$ of possible saccade commands. Following these definitions, the objective of the "Where" processing pathway is to allow a saccadic decision by training such a likelihood function $w(u|\mathbf{s}^t)$ from observing failures and success from different saccades selection.

Now, the optimization being done on u, our saccade selection process relies on maximizing the likelihood of success, i.e. $\arg\max_u p(\mathbf{o} = 1|u, \mathbf{s}^t)$, that is consistent with assuming that a prior is put on observing a success, whatever the saccade. Computing a good approximation of the likelihood $p(\mathbf{o} = 1|u, \mathbf{s}^t)$ is therefore crucial to perform visual search:

$$w(u|\mathbf{s}^t) \approx p(\mathbf{o} = 1|u, \mathbf{s}^t) = A^t(\rho(u)) \tag{3}$$

where $\rho(u)$ is the future position of gaze for a saccade u, and $\mathbf{s}^t$ is the feature vector representing the present peripheral observation. The model predicts the accuracy of the "What" pathway, given the action u (saccade).

The choice of a saccade given the likelihood may be obtained from the maximum a posteriori rule :

$$\pi_{\max}(\mathbf{s}^t) = \arg\max_u p(\mathbf{o} = 1|u, \mathbf{s}^t) \cdot \Pr(u) \tag{4}$$

With for instance $\Pr(u) = \text{Unif}(u)$ a uniform prior probability on saccade selection, that is, uniformly on motor space, we obtain the policy (approximate in probability):

$$\pi_{\max}(\mathbf{s}^t) \approx \hat{\pi}_{\max}(\mathbf{s}^t) \stackrel{\text{def.}}{=} \arg\max_u w(u|\mathbf{s}^t) \tag{5}$$

Similarly, another strategy would be to use the approximate conditional expectation on action space:

$$\hat{\pi}_{\text{avg}}(\mathbf{s}^t) \stackrel{\text{def.}}{=} \int_u u \cdot w(u|\mathbf{s}^t) \cdot \Pr(u) \cdot du \tag{6}$$

Note that this conditional expectation is different from that that would operate in cartesian coordinates. In particular, using a log-polar accuracy map comes with an intrinsic prior for the saccades to be closer to the fixation point (see Fig. 2).

Incidentally, the unimodal shape of the accuracy map indicates that a highest chance of success is found when the target is centered on the fovea, and for that reason the active inference mechanism should privilege saccades that will place the visual target at the center of the fovea. This is equivalent to identifying the location of the target in the retinotopic space, and thus inferring the spatial information from the visual field, with the future saccade taking the role of a latent variable explaining the current visual field $\mathbf{s}^t$.

From the active inference perspective, choosing the accuracy map as a likelihood function is like putting a prior on observing a success. In other words, the agent is more "surprised" in case of classification failure than in case of classification success. Taking the classification success as the principal outcome of the "Where" pathway, the action selection process now relies on minimizing the surprise as upper-bounded by the free-energy:

$$-\log p(\mathbf{o} = 1|\mathbf{s}^t) \leq F \text{ with } F \overset{\text{def.}}{=} \mathbb{E}_q[-\log p(\mathbf{o} = 1|\pi, \mathbf{s}^t) + \log q(\pi|\mathbf{s}^t, \mathbf{o} = 1) - \log p(\pi|\mathbf{s}^t)] \tag{7}$$

with π the policy taking the role of a latent variable predicting the (future) classification success, and q being a probability distribution function on action selection policy. Finally, the visual search problem can be summarized as optimizing the function q which would define a saccade selection policy from a maximum success evidence perspective.

3.3   Higher Level Inference: Choosing the Processing Pathway

Inferring the target location and identity sums up in our case to select a saccade in order to infer the target category from the future visual field. It is likely, however, that a saccade may not provide the expected visual data, and that a corrective saccade may be needed to improve the visual recognition. More generally, choosing to move the eye or to issue a categorical response from the available data resorts to select one processing pathway over the other: either realize the saccade or guess the category from the current foveal data. In order to make this choice, one must guess whether the chance of success is higher in the present, given the current visual field, or in the future, after the next saccade.

This, again, can be expressed under the active inference setup. Let $p(\mathbf{o}|f(x_0))$ the probability of success when processing the foveal data, as provided by the "What" network. Under the policy π (provided by the "Where" network), the decision decomposes into a binary choice between issuing a saccade or not. This decision should rely on comparing $p(\mathbf{o}|f(\rho(\pi(\mathbf{s}))$ (the future accuracy) and $p(\mathbf{o}|f(x_0))$ (the current accuracy). The active inference comes down here to a binary choice between actuating a saccade or "actuating" (testing) the categorical response.

Interestingly, the log difference of the two probabilities

$$\log p(\mathbf{o}|f^t(\rho(\pi(\mathbf{s}^t)) - \log p(\mathbf{o}|f^t(x_0)) \sim \log A^t(\rho(\pi(\mathbf{s}^t))) - \log A^t(x_0) \tag{8}$$

can be seen as an estimator of the *information gain* provided by the saccade. Choosing to actuate a saccade is thus equivalent to maximising the information gain provided by the new visual data, consistently with the classic "Bayesian surprise" metric [19]. Expanding over purely phenomenological models, our model finally provides a biologically interpretation of the information gain metric as a high-level decision criterion, linked to the comparison of the output of the two principal visual processing pathways.

3.4 Learning the Accuracy Map

Neural Networks are known to be in theory universal value function approximators and in practice, we will use a network architecture, alike to that used for the "What" pathway. This will provide a sufficient argument for showing that it is possible to learn such a mapping, while leaving open the possibility that other architectures may be actually implemented in the brain. The parametric neural network consists of the input feature vector $\mathbf{s}^t$ (of dimension 2880), followed by two fully-connected hidden layers of size 1000 with rectified linear activation units (ReLUs). A final fully-connected output layer with a sigmoid nonlinearity ensures that the output is compatible with a likelihood function. In accordance with observations [18,20], the same log-polar compression pattern is defined at the retinal input and at the motor output (see Fig. 1).

To learn the mapping provided by the "Where" network, we use the BCE cost as the Kullback-Leibler divergence between the tested accuracy and its approximation:

$$\mathcal{L}_S^t = -[\mathbf{o}(f^t(\rho(u^t))) \cdot \log w(u^t|\mathbf{s}^t) + (1 - \mathbf{o}(f^t(\rho(u^t)))) \cdot \log(1 - w(u^t|\mathbf{s}^t))] \quad (9)$$

We then optimize the parameters of the neural network implementing the "Where" pathway such as to optimize the approximation of the likelihood function. This can be achieved in our feed-forward model using back-propagation [2] with the input-output pairs $(\mathbf{s}^t, u^t)$ and the classification result as it is given by the "What" pathway. The role of the "What" pathway is here that of a critic of the output of the "Where" pathway (which takes the role of the actor). This separation of visuo-spatial processing into an actor and a critic is reminiscent of a more general actor-critic organization of motor learning in the brain, as postulated by Joel, Niv, and Ruppin [21].

The natural way to collect such supervision data is to draw data one by one in our virtual experiment, iteratively generating a saccade and computing the success of the detection. This is what would be performed by an agent which would sequentially learn by trial-and-error, using the actual recognition accuracy (after the saccade) to grade the action selection and leading to a reinforcement scheme. For instance, we could use corrective saccades to compute (a posteriori) the probability of a correct localization. In a computer simulation however, this calculation is slow and not amenable. To accelerate the learning in our scheme defined by a synthetic generative model, there exists however a computational shortcut to obtain more supervision pairs. Indeed, the learning of the where pathway may be done after that of the what pathway. Such a computational shortcut is allowed by the independence of the categorical performance with position. Moreover, for each input image, we know the true position in extrinsic (x^t) and intrinsic $(u^t \stackrel{\text{def}}{=} \rho^{-1}(x^t))$ and identify k^t of the target. As such, one can compute the average accuracy map over the dataset and optimize equivalently

$$\mathcal{L}_{\mathcal{S}}^t = - \sum_{u \in \mathcal{S}} [A_0(u - u^t) \cdot \log w(u|\mathbf{s}^t) + (1 - A_0(u - u^t)) \cdot \log(1 - w(u|\mathbf{s}^t)] \quad (10)$$

where $A_0(\Delta u)$ stands for the accuracy map with respect to the true position, that is, of the accuracy when the input image to the "What" pathway is systematically shifted by $\rho(\Delta u)$. In our setting, this function varied little for different identities and we averaged it over all possible identities. Combining this translational shift and the shift-dependent accuracy map of the "What" classifier, the actual accuracy map at each trial can be thus predicted under an ergodic assumption by shifting the central accuracy map on the true position of the target (that is with $\Delta x = \rho(u) - x^t$). Then, this full accuracy map is a probability distribution function which can be computed on the rectangular grid of the visual display. We project this distribution on a log-polar grid to provide the expected accuracy of each hypothetical saccade in a retinotopic space similar to a collicular map. Applied to the full sized ground truth accuracy map computed in metric space, this gives an accuracy map at the different positions of the retinotopic motor space $\mathcal{S}$. This accelerate learning as it scales up both the set of tested saccade positions and gives the analog bias value instead of the binary outcome of the detection. Future work should explore if similar results will still hold when both networks are learned at the same time and with a trial-and-error strategy.

4   Results

After training, we observed that the "Where" pathway can correctly predict an accuracy map, whose maximal argument can be chosen to drive the eye toward a new viewpoint with a single saccade. There, a central snippet is extracted, that is processed through the "What" pathway, allowing to predict the digit's label. The full scripts for reproducing the figures and explore the results to the full range of parameters is available at https://github.com/laurentperrinet/WhereIsMyMNIST (under a GPLv3 license). The network is trained on 60 epochs of 60000 samples, with a learning rate equal to 10^{-4} and the Adam optimizer [22] with standard momentum parameters. An improvement in convergence speed was obtained by using batch normalization. One full training takes about 1 hour on a laptop. The code is written in Python (version 3.7.6) with the pyTorch library [23] (version 1.1.0).

Saccades distributions and classification success statistics resulting from this simple sequence are presented in Fig. 2. Figure 2A-C provides an example of our active visual processing setup. The initial visual field (Fig. 2A) is processed through the "Where" pathway, providing a predicted accuracy map (compared with the true accuracy map in Fig. 2B)). The maximal argument of the accuracy map allows to actuate a saccade. The resulting visual field is provided in

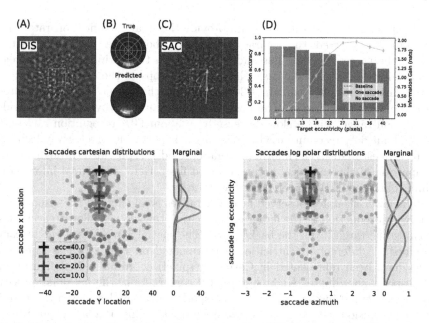

Fig. 2. Example of active vision after training the "Where" network. Digit contrast set to 70%. From left to right: **(A)** Magnified reconstruction of the visual input, as reconstructed from the primary visual feature vector through an inverse log-polar transform. **(B)** Color-coded radial representation of the output accuracy maps, with dark blue for the lower accuracy values, and yellow for higher values. The network output ("Predicted") is visually compared with the ground truth ("True"). **(C)** Visual field shift obtained after doing a saccade: The digit (the number 4) can now be recognized within the foveal region. **(D)** The final classification rate is plotted in function of the target eccentricity. The transparent orange corresponds to the pre-saccadic accuracy from the central classifier ('no saccade'). The blue bars correspond to the post-saccadic accuracy ('one saccade'), averaged over 1000 trials per eccentricity. Red line: empirical information gain, estimated from the accuracy difference. **(E)** Saccades distribution for different target eccentricities. The same saccades are plotted in (pixel) Cartesian coordinates on the left, and in log-polar coordinates on the right. The Cartesian coordinates correspond to the effector space while the log-polar coordinates correspond to the motor control space. In both cases, the empirical marginal distributions over one axis are shown on the right side. (Color figure online)

Fig. 2C, and the classification is done on the central part of the visual field only (red square). To generalize results, 1000 saccades are sampled for different sequences of input visual fields containing a target with a fixed eccentricity, but a variable identity, a variable azimuth and a variable background clutter. The digit contrast parameter is set to 70% and the eccentricity varies between 4 and 40 pixels. The empirical classification accuracies are provided in Fig. 2D, for different eccentricities. These are averaged over all trials both on the initial central snippet and the final central snippet (that is, at the landing of the saccade). The (transparent) orange bars provide the initial classification rate (without saccade) and the blue bars provide the final classification rate (after saccade). As expected, the accuracy decreases in both cases with the eccentricity, for the targets become less and less visible in the periphery. The decrease is rapid in the pre-saccadic case: the accuracy drops to the baseline level for a target distance of approximately 20 pixels from the center of gaze, consistent with the size of the target. The post-saccadic accuracy provides a much wider recognition range, with a slow decrease from about 90% recognition rate up to up to about 60% recognition when the target is put at 40 pixels away from the center. An estimate of the information gain provided is provided through a direct comparison of the empirical accuracies (red line). Here an optimal information gain is obtained in the 25–35 eccentricity range.

The lower accuracy observed at larger ranges is an effect of the visual signal bandwidth reduction at the larger eccentricities, that do not allow to accurately separate the target from the background. The spatial spreading of the saccades obtained at different eccentricities is represented on Fig. 2E. The same saccades have been represented in Cartesian (pixel) coordinates (left figure) and in log-polar coordinates (right figure). By construction, the log-polar processing, implemented in the "Where" visuo-spatial pathway, leads to a decrease in saccade precision with respect to the eccentricity. This decreasing precision is illustrated by the higher variance of the saccades distribution observed at higher eccentricities, in the Cartesian space of the saccade realization. Interestingly, the variance of the marginal distribution of the saccades along the eccentricity axis is close to constant when represented in the log-polar space, that is, in the space of the (collicular) motor command. From 10 to 30 pixels away from the center, the precision of the command is invariant with respect to the eccentricity. The lower precision observed at about 40 pixels eccentricity only reflects a lower detection rate. Due to the log-polar construction of the motor map, the motor command (falsely) appears to display the same precision at various eccentricities. As it would be the case with a more detailed model of the motor noise, this log-polar organization of the control space can be interpreted as a natural renormalization, helping to counteract the precision loading that would otherwise be attached with the larger saccades, helping to provide a more uniform spread of the motor command in the effector space.

5   Discussion and Perspectives

We proposed a computer-based framework allowing to implement visual search under bio-realistic constraints, using a foveated retina and a log-polar visuo-motor control map. A simple "Naïve Bayes" assumption justifies the separation of the processing in two pathway, the "What" visuo-semantic pathway and the "Where" visuo-spatial pathway. The predicted classification rate (or classification accuracy), serves as a guiding principle throughout the paper. It provides a way to link and compare the output of both pathways, serving either to select a saccade, in order to improve the chance of success, or to test a categorical response on the current visual data.

Future work should explore the application of this architecture to more complex tasks, and in particular to a more ecological virtual experiment consisting in classifying natural images. In particular, it would be possible to generalize this to a sequence of saccades, that is, mapping out an entire sequence of saccades by the where pathway, given the current field of view [24]. Finally, we used here the log-polar retinotopic mapping as a constraint originating from the anatomy of the visual pathways and have shown in Fig. 2 that this implicitly generate a uniform action selection probability. At the temporal scale of natural selection, one could also consider this mapping as the emergence of an optimal solution considering an ecological niche, explaining for instance why foveal regions are more concentrated in predators than in preys, as shown for instance in avians [25]. As can be observed in the comparative study of pupils' shapes [26], this may justify the differences observed between preys (with a less sparse cone density at the periphery) and predators (with a tendency toward denser foveal regions) as a form. The compromise between the urgency to detect and the need to be accurate may justify the different balances which may exist in different species and thus as long term form of homeostasis [27].

References

1. Sanz-Leon, P., Vanzetta, I., Masson, G., Perrinet, L.U.: Motion clouds: model-based stimulus synthesis of natural-like random textures for the study of motion perception. J. Neurophysiol. **107**(11), 3217–3226 (2012). https://doi.org/10.1152/jn.00737.2011
2. Lecun, Y., Bottou, L., Bengio, Y., Haner, P.: Gradient-based learning applied to document recognition. Proc. IEEE **86**(11), 2278–2324 (1998). http://doi.org/10/d89c25
3. Tang, Y., Nguyen, D., Ha, D.: Neuroevolution of self-interpretable agents. In: Proceedings of the 2020 Genetic and Evolutionary Computation Conference (2020). http://doi.org/10/gg64b3
4. Friston, K.J., Adams, R.A., Perrinet, L.U., Breakspear, M.: Perceptions as hypotheses: saccades as experiments. Front. Psychol. **3** (2012). https://doi.org/10.3389/fpsyg.2012.00151
5. Daucé, E.: Active fovea-based vision through computationally-effective model based prediction. Front. Neurorobotics **12** (2018). http://doi.org/10/gfrhgj
6. Friston, K.: The free-energy principle: a unified brain theory? Nat. Rev. Neurosci. **11**(2), 127–138 (2010)
7. Perrinet, L.U., Adams, R.A., Friston, K.J.: Active inference, eye movements and oculomotor delays. Biol. Cybern. **108**(6), 777–801 (2014). https://doi.org/10.1007/s00422-014-0620-8
8. Mirza, M.B., Adams, R.A., Mathys, C., Friston, K.J.: Human visual exploration reduces uncertainty about the sensed world. PLoS ONE **13**(1), e0190429 (2018). https://doi.org/10.1371/journal.pone.0190429
9. Cullen, M., Monney, J., Mirza, M.B., Moran, R.: A Meta-Bayesian Model of Intentional Visual Search (2020)
10. Treisman, A.M., and Gelade, G.: A feature-integration theory of attention. Cogn. Psychol. **12**(1), 97–136, 11957 (1980). https://doi.org/10.1016/0010-0285(80)90005-5

11. Eckstein, M.P.: Visual search: a retrospective. J. Vision **11**(5), 14–14 (2011). http://doi.org/10/fx9zd9
12. Wolfe, J.M.: Visual search. In: The Handbook of Attention, pp. 27–56. MIT Press, Cambridge(2015)
13. Najemnik, J., Geisler, W.S.: Optimal eye movement strategies in visual search. Nature **434**(7031), 387–391 (2005). http://doi.org/10/bcbw2b
14. Daucé, E., Albiges, P., Perrinet, L.U.: A dual foveal-peripheral visual processing model implements efficient saccade selection. J. Vis. **20**(8), 22–22 (2020). https://doi.org/10.1167/jov.20.8.22
15. Watson, A.B.: A formula for human retinal ganglion cell receptive field density as a function of visual field location. J. Vis. **14**(7), 15–15 (2014)
16. Javier Traver, V., Bernardino, A.: A review of log-polar imaging for visual perception in robotics. Robot. Auton. Syst. **58**(4), 378–398 (2010)
17. Fischer, S., Sroubek, F., Perrinet, L.U., Redondo, R., Cristóbal, G.: Selfinvertible 2D log-Gabor wavelets. Int. J. Comput. Vis. **75**(2), 231–246 (2007). https://doi.org/10.1007/s11263-006-0026-8
18. Sparks, D.L., Nelson, I.S.: Sensory and motor maps in the mammalian superior colliculus. Trends Neurosci. **10**(8), 312–317 (1987)
19. Itti, L., Baldi, P.: Bayesian surprise attracts human attention. Vis. Res. **49**(10), 1295–1306 (2009). https://doi.org/10.1016/j.visres.2008.09.007
20. Connolly, M., Van Essen, D.: The representation of the visual field in parvocellular and magnocellular layers of the lateral geniculate nucleus in the macaque monkey. J. Comp. Neurol. **226**(4), 544–564 (1984)
21. Joel, D., Niv, Y., Ruppin, E.: Actor-critic models of the basal ganglia: new anatomical and computational perspectives. Neural Netw. **15**(4–6), 535–547 (2002)
22. Kingma, D.P., and Ba, J.: Adam: a method for stochastic optimization. arXiv preprint arXiv:1412.6980 (2014)
23. Paszke, A., et al.: PyTorch: an imperative style, high-performance deep learning library. In: Wallach, H., Larochelle, H., Beygelzimer, A., dAlché-Buc, F., Fox, E., Garnett, R. (eds.) Advances in Neural Information Processing Systems, vol. 32, pp. 8024–8035. Curran Associates, Inc.(2019)
24. Hoppe, D., and Rothkopf, C.A.: Multi-step planning of eye movements in visual search. Sci. Rep. **9**(1), 144 (2019). http://doi.org/10/gfwcvc
25. Moore, B.A., Tyrrell, L.P., Pita, D., Bininda-Emonds, O.R.P., Fernández-Juricic, E.: Does retinal configuration make the head and eyes of foveate birds move? Sci. Rep. **7** (2017). http://doi.org/10/f9k78h
26. Banks, M.S., Sprague, W.W., Schmoll, J., Parnell, J.A.Q., and Love, G.D.: Why do animal eyes have pupils of different shapes? Sci. Adv. **1**(7), e1500391 (2015). http://doi.org/10/gg66t6
27. Connant, R.C., and Ashby, W.R.: Every good regulator of a system must be a model of that system. Int. J. Syst. Sci. **1**(2), 89–97 (1970). http://doi.org/10/bbgr9b

Sophisticated Affective Inference: Simulating Anticipatory Affective Dynamics of Imagining Future Events

Casper Hesp[1,2,3,4(✉)], Alexander Tschantz[5], Beren Millidge[6], Maxwell Ramstead[6,7,8], Karl Friston[4], and Ryan Smith[9]

[1] Department of Psychology, University of Amsterdam, Amsterdam, The Netherlands
c.hesp@uva.nl
[2] Amsterdam Brain and Cognition Centre,
University of Amsterdam, Amsterdam, The Netherlands
[3] Institute for Advanced Study, University of Amsterdam, Amsterdam, The Netherlands
[4] Wellcome Centre for Human Neuroimaging, University College London, London, UK
[5] Sackler Centre for Consciousness Science, School of Engineering and Informatics,
University of Sussex, Sussex, UK
[6] School of Informatics, University of Edinburgh, Edinburgh, UK
[7] Division of Social and Transcultural Psychiatry, Department of Psychiatry, McGill University,
Montreal, Canada
[8] Culture Mind and Brain Program, McGill University, Montreal, Canada
[9] Laureate Institute for Brain Research, Tulsa, OK, USA

Abstract. In this paper, we combine sophisticated and deep-parametric active inference to create an agent whose affective states change as a consequence of its Bayesian beliefs about how possible future outcomes will affect future beliefs. To achieve this, we augment Markov Decision Processes with a Bayes-adaptive deep-temporal tree search that is guided by a free energy functional which recursively scores counterfactual futures. Our model reproduces the common phenomenon of rumination over a situation until unlikely, yet aversive and arousing situations emerge in one's imagination. As a proof of concept, we show how certain hyperparameters give rise to neurocognitive dynamics that characterise imagination-induced anxiety.

Keywords: Affect · Counterfactuals · Anxiety · Active inference · Anticipation

1 Introduction

A common aspect of human experience is that imagined, counterfactual events can have a significant impact on our affective states. In its extreme form, people suffering from a variety of psychiatric conditions, such as generalised anxiety disorder (Gale and Davidson 2007), consistently report experiencing repetitively imagined "what-if" scenarios that have a significant impact on their real-time affective dynamics. This type of maladaptive, repetitive thinking about (often unlikely) negative future outcomes is referred

© Springer Nature Switzerland AG 2020
T. Verbelen et al. (Eds.): IWAI 2020, CCIS 1326, pp. 179–186, 2020.
https://doi.org/10.1007/978-3-030-64919-7_18

to as rumination. Clinically validated therapeutic interventions for disorders involving rumination (e.g., cognitive-behavioural therapy [CBT], acceptance and commitment therapy [ACT]) also typically aim to reduce confidence in catastrophic imagined future events and ground patients in the here and now (e.g., see Barlow et al. 2017; Hayes et al. 2006). Although the effectiveness of such therapies is well established, their mechanisms of action remain poorly understood. Gaining a more detailed understanding of the specific neurocomputational mechanisms that underpin prospection-induced affect in general – and excessive rumination-induced anxiety in particular – is an important direction for future research.

In this paper, we aim to provide a mechanistic account of how affective responses can be generated by imagined future outcomes – and how this can become dysfunctional during rumination. By combining two recent developments in active inference, we provide a formal model of these phenomena and simulate how 'overthinking a situation' can occur – continuing to the point where unlikely, yet aversive and arousing situations emerge in one's imagination. We employ an affective-inference agent (Hesp et al. 2020) equipped with the recursive belief-updating scheme of sophisticated inference (Friston et al. 2020). This powerful combination allows us – for the first time – to create an agent whose affective states change as a consequence of its internal machinations about possible future events. In this short paper, we present the underlying generative model and discuss its implications. We also show some brief illustrative simulations. We leave a more elaborate analysis of computational results for a variety of parametrisations for a future piece.

2   Methods

Here, we show how one can augment the Markov Decision Process formalism that underwrites the standard active inference scheme with a Bayes-adaptive deep-temporal tree search that is guided by a free energy functional as it scores counterfactual futures. By combining the ensuing recursive update scheme of sophisticated inference (Friston et al. 2020) with deep-parametric, *affective* inference (Hesp et al. 2020), we can derive a general-purpose generative model of the following mathematical form, summarised graphically in Fig. 1 and in tabular format in Table 1:

$$P\left(\tilde{o}, \tilde{s}^{(1)}, \tilde{u}\tilde{\gamma}, , s^{(2)}\right) =$$

$$P\left(o_1|s_1^{(1)}\right)P\left(s_1^{(1)}|s^{(2)}\right)P\left(s^{(2)}\right)\prod_{\tau=1}^{T-1}P\left(o_{\tau+1}|s_{\tau+1}^{(1)}\right)P\left(s_{\tau+1}^{(1)}|s_\tau^{(1)}, u_\tau\right)P\left(u_\tau|\gamma_\tau, s^{(2)}\right)P\left(\gamma_\tau|s^{(2)}\right) \quad (1)$$

In brief, the (higher-level) affective-contextual states $s^{(2)}$ entail three hidden-state factors: arousal, valence, and context. These factors map (through the likelihood matrix $\mathbf{A}^{(2)}$) onto three lower-level model variables: the latent states $s_1^{(1)}$, actions u_τ (i.e., possible state transitions at time τ), and $\mathbf{G}_\tau$-precision γ_τ (i.e., action confidence at time τ). The latter is a scalar precision that scales the contribution of the expected free energy $\mathbf{G}_\tau$ to posterior beliefs about actions. This precision term can be read as a subjective estimate of confidence in model-based beliefs about action outcomes (Hesp et al. 2020). This estimate is updated when posterior beliefs about action depart from one's prior expectations such that it produces a con-comitant change in the action-averaged expected free

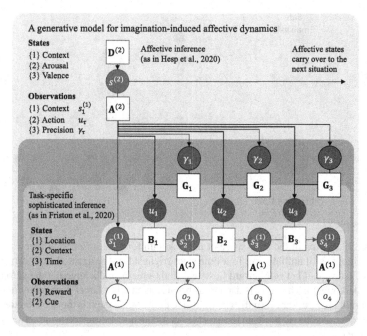

Fig. 1. A directed acyclic Bayes graph showing a generative model for sophisticated affective inference about higher-level valence, arousal and context states ($s^{(2)}$) based on (imagined) lower-level action-model precision (γ_τ), actions (u_τ), states ($s_\tau^{(1)}$), and outcomes (o_τ) over four successive time points, thus combining sophisticated active inference (Friston et al. 2020) with deep-parametric affective inference (Hesp et al. 2020).

energy. The ensuing update term—named "affective charge" or *AC*—reflects changes in the confidence in one's action model.

The lower-level state space $\tilde{s}^{(1)}$ comprises three hidden-state factors: location, context, and time, which map (through $\mathbf{A}^{(1)}$) onto two outcome modalities representing cues (e.g., visual) and rewards (e.g., gustatory). Following Hesp et al. (2020), each of the higher-level states can be associated with different combinations of lower-level parameters for $s_1^{(1)}$ (in terms of the initial prior $\mathbf{D}^{(1)}$), u_t (in terms of the baseline action prior $\mathbf{E}_t$), and γ_t (in terms of the rate parameter $\boldsymbol{\beta}_\tau$) through a higher-level likelihood mapping $\mathbf{A}^{(2)}$. For example, imagine you experience a pleasant low arousal state when you arrive home after a day's hard work. This higher-level belief about your current state can then inform your lower-level action beliefs, e.g., by increasing the prior probability of actions associated with getting ready to sleep. Conversely, imagining yourself getting ready to sleep can further increase your experienced sleepiness. It is the latter type of reaction that we would like to model in general: affective responses (in this case, arousal-reducing responses) generated by imagined (internally simulated) future events.

The specifics of the lower level generative model are not terribly important, but for the sake of our demonstration, we introduce a simple state space (within a stable context) that comprises four states (see Fig. 2), each with its own observable outcome and associated preference **C**:

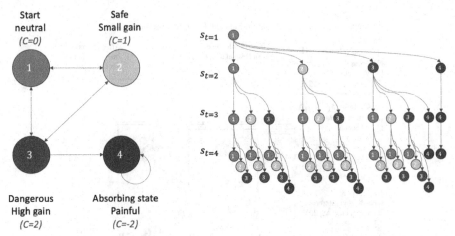

Fig. 2. An illustration of the state-space of the task with four states (left side, arrows indicating likely transitions) as it unfolds over four time steps (right side). The agent always starts in state 1 (grey), can get a small but safe reward in state 2 (light green), and a large but dangerous reward in state 3 (dark green). The latter is dangerous because it entails a larger probability of transition to the absorbing painful state 4 (red). The right side of the figure depicts the decision tree through which the agent searches to evaluate the expected consequences of each possible action sequence. (Color figure online)

state 1: an initial neutral state $(\mathbf{C}_{s=1} = 0)$ e.g., being at home base
state 2: a slightly rewarding state $(\mathbf{C}_{s=2} = +1)$ e.g., picking berries
state 3: a highly rewarding state $(\mathbf{C}_{s=3} = +2)$ e.g., hunting large prey
state 4: a painful absorbing state $(\mathbf{C}_{s=4} = -2)$. e.g., being wounded

The agent always starts in neutral state 1 and can move towards any of the four states by selecting up to three moves. Furthermore, a notion of safety is introduced by making state 2 a safer option than state 3: transitioning towards the latter has a higher probability of failure and can accidentally lead to painful *state 4*, which cannot be left until the end of the (4 time-step) trial. If we liken a trial to a working day, state 1 could be seen as the agent's home base, state 2 as a safe activity with a small yet certain reward (e.g., picking berries), state 3 as a dangerous activity with a large yet uncertain reward (e.g., hunting prey), and state 4 as an unpreferred state that cannot be left for the rest of the day (e.g., being wounded).

An important twist introduced in this model is that higher-level state beliefs can be updated recursively through pre-task mental deliberation, based on a deep tree search that unfolds pre-emptively on the lower-level. All the equations presented in Table 1 can be evaluated without presenting any actual outcomes to the agent in question – that is, belief updating is guided by the probabilistic exploration of possible futures. This tree search involves recursive updating of lower-level action beliefs based on the counterfactual outcomes of actions that are sampled from predictive posteriors at each branching point of the tree. Because we equip the generative model with action-dependent $\mathbf{G}_\tau$-precision estimation, we can see how each counterfactual future elicits an *expected affective charge* (*eAC*; see the first row of Table 1), which provides an ascending message to inform

Table 1. This table lists the predictive posteriors that provide the empirical priors for our generative model.

Predictive posteriors	Mathematical definitions	
$Q(s^{(2)}) = Cat(\mathbf{s}^{(2)})$	$\mathbf{s}^{(2)} = \sigma[\ln \mathbf{D}^{(2)}$	*(higher-level state prior)*
higher-level state beliefs:	$+\sum_\tau \ln \mathbf{A}_{\mathbf{E}_\tau} \cdot \mathbf{u}_\tau$	*(action evidence)*
context, arousal, valence	$-\sum_\tau \left(\ln \frac{\beta^{(+,-)} - eAC_\tau}{\beta^{(+,-)}} \cdot \frac{\beta_\tau}{\beta_\tau - eAC_\tau}\right)]$	
		(affective evidence)
	$eAC_\tau = (\mathbf{u}_\tau^o - u_\tau) \cdot \mathbf{G}(u_\tau, o_\tau)$	*(expected affective charge)*
$Q(s_\tau^{(1)}) = Cat(\mathbf{s}_\tau^{(1)})$	$\mathbf{s}_1^{(1)} = \mathbf{D}^{(1)} = \mathbf{A}_s^{(2)}\mathbf{s}^{(2)}$	*(lower-level initial state prior)*
lower-level state beliefs:	$\mathbf{s}_\tau^{(1)} \propto (\ln \mathbf{A} \cdot o_\tau) \odot \mathbf{s}_\tau^u$	*(lower-level empirical state prior)*
context, location, time		
$Q(s_\tau^{(1)}|u_{s\tau}) = Cat(\mathbf{s}_{\tau+1}^{(1)u})$	$\mathbf{s}_{\tau+1}^u = \mathbf{B}(u_\tau)\mathbf{s}_\tau$	*(action-dependent state priors)*
action-specific state expectations		
$Q(u_\tau|o_\tau, \gamma_\tau, s^{(2)}) = Cat(\mathbf{u}_\tau^o)$	$\mathbf{u}_\tau^o = \sigma[\mathbf{E}_\tau + \gamma_\tau \mathbf{G}(u_\tau, o_\tau)]$	*(full action prior)*
outcome- and time-specific	$\mathbf{E}_\tau = \mathbf{A}_{\mathbf{E}_\tau}^{(2)}\mathbf{s}^{(2)}$	*(baseline action prior)*
action expectations	$\gamma_\tau = 1/\beta_\tau$	*(time-specific expected $\mathbf{G}$-precision)*
based on higher-level states	$\mathbf{G}_\tau^{u,o} = \mathbf{G}(u_\tau, o_\tau) =$	*(path-specific expected free energy)*
and expected $\mathbf{G}_\tau$-precision	$\mathbf{o}_{\tau+1}^u \cdot (\ln \mathbf{o}_{\tau+1}^u + \mathbf{C})$	*(expected risk)*
	$+s_{\tau+1}^u \cdot \mathbf{H}$	*(expected ambiguity)*
	$+\mathbf{u}_{\tau+1}^o \cdot \mathbf{G}_{\tau+1}^{u,o}\mathbf{o}_{\tau+1}^u$	*(expected free energy of subsequent actions)*
	$\mathbf{C} = \ln P(o_\tau)$	*(prior preferences)*
$Q(o_\tau|u_{<\tau}) = Cat(\mathbf{o}_\tau^u)$	$\mathbf{o}_\tau^u = \mathbf{A}\mathbf{s}_\tau^u$	
action-specific		
outcome expectations		
$Q(\gamma_\tau|s^{(2)}) = \Gamma(1, \boldsymbol{\beta}_\tau)$	$\boldsymbol{\beta}_\tau = \beta^{(+,-)} \cdot \mathbf{A}_\beta^{(2)}\mathbf{s}^{(2)}$	*(expected rate parameter)*
time-specific $\mathbf{G}_\tau$-precision		
based on higher-level states		

higher-level affective inference. The equation for *eAC* deserves further unpacking:

$$eAC_\tau = \left(\mathbf{u}_\tau^o - u_\tau^o\right) \cdot \mathbf{G}(u_\tau, o_\tau) \tag{2}$$

Where $\mathbf{u}_\tau^o$ is the empirical prior for a particular action and outcome at time τ, and u_τ^o is a particular outcome-action sequence drawn from the predictive posteriors. The eAC_τ term thus scores imagined departures from the model-averaged expected free energy for an imagined future at time τ. This *eAC* term is the anticipatory analogue of the affective charge term proposed by Hesp and colleagues (2020) as a plausible source of evidence for different valence states (i.e., pleasant/unpleasant states). The two main innovations afforded by sophisticated inference are that: (i) in e *AC*, the action sequences consider all combinations of individual actions and (ii) *eAC* is elicited in response to imagined, counterfactual actions (as opposed to events that have already been observed).

Simulating all possible sequences of actions and outcomes would quickly become intractable due to a combinatorial explosion (right side of Fig. 2). For example, with 4 possible outcomes, actions, and time steps, the number of imaginable future possibilities would exceed 16,000. To solve this problem, sophisticated inference (Friston et al. 2020) provides a principled way of exploring the tree using the certainty of predictive posteriors. In terms of state estimation, these can be seen as empirical priors – as they are derived entirely from prior beliefs, which inform sampling of possible futures. In this work, every path has a probability of being selected, however small. Obviously, the number of explored possibilities will tend to increase with each iteration. By manipulating the number of iterations of such self-directed, recursive sampling of the future we can model traditional speed-accuracy tradeoffs for split-second decisions (i.e., too few iterations) as well as the detrimental effects of excessive deliberation (i.e., too many iterations), which characterises the phenomenon of rumination or 'overthinking'.

3   Results

An exemplar result from our simulations is shown in Fig. 3 below. It provides a simple demonstration of how sophisticated affective inference naturally underwrites affective responses to internally imagined futures. We simulated how particular hyperparameters give rise to neurocognitive dynamics that characterise imagination-induced anxiety or pessimism about the future. In particular, in Fig. 3 we show how iterating the tree search too often (i.e., 'overthinking') can trigger recursive reductions in $\mathbf{G}_\tau$-precision as the agent enters the following vicious cycle: (1) Every time they end up imagining a very negative outcome, their action-model confidence is reduced. (2) Every reduction in expected precision γ_τ (for simplicity assumed to be the same for all τ) will reduce reliance on one's action-model for subsequent explorations of the future because these are sampled from the predictive posterior over action (see the fourth entry of Table 1). This type of excessive, negatively biased prospection (i.e., rumination) will subsequently influence the higher-level affective state, which recursively affects the lower-level sampling algorithm in multiple ways.

Crucially for these simulations of rumination, a negative affective state can bias the agent's expectations towards negative outcomes and reduce lower-level $\mathbf{G}_\tau$-precision even further, leading to increasingly pessimistic exploration of the tree. Such affective decision-tree pruning has been observed in a number of previous studies (Dayan and Huys 2008; Huys et al. 2012; Huys et al. 2015; Níally et al. 2017). Our work shows how this phenomenon can be cast as a form of belief-updating under sophisticated affective inference (Hesp et al. 2020; Friston et al. 2020). Furthermore, the aetiology of many other psychiatric conditions seems to be intimately related to affective responses to imagined events: cravings in addiction, intrusive thoughts in obsessive-compulsive

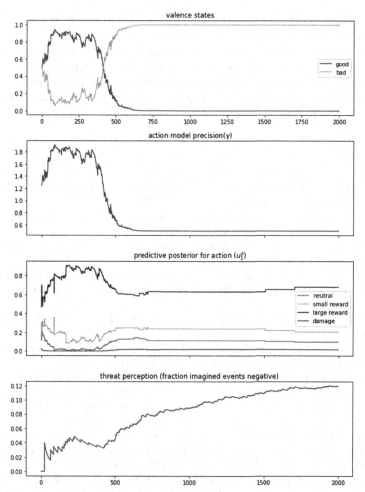

Fig. 3. An example of simulation results showing detrimental effects of overthinking (i.e., rumination) when considering affective responses to imagined future events. Horizontal axes indicate the number of iterations and, implicitly, the amount of time allowed for internal deliberation. The top panel shows Bayesian beliefs about good and bad valence states (blue and orang, respectively); the second panel shows expected precision (blue); the third panel shows the predictive posterior for each possible first action: moving to either the neutral location (grey), the small reward (light green), the large one (dark green), or the painful absorbing state (red); the bottom panel shows the fraction of imagined events that were negative. Initially, exploration gives rise to an optimistic phase of increasingly positive valence (blue line in top panel), increasing action-model precision (second panel), increasingly positive expectations about future state transitions (dark green line in third panel) and a relatively small fraction of imagined negative events (red line in bottom panel). However, after roughly 500 iterations of the deep tree search, the agent devolves into a state of negative affect, reduced action-model precision, pessimistic expectations about future rewards, and a much higher fraction of imagined negative events. (Color figure online)

disorder, flashbacks in post-traumatic stress disorder, hallucinations and delusions in schizophrenia, fear of gaining weight in anorexia, excessive monitoring of self-states in anxiety, and so forth. As such, this type of formal model of imagination-induced affective responses could represent an important step forward in computational psychiatry and might one day be extended to aid in diagnosis or treatment for a variety of affective disorders – thus working towards computational nosology and precision psychiatry (see Friston et al. 2017).

References

Barlow, D.H., et al.: Unified Protocol for Transdiagnostic Treatment of Emotional Disorders: Therapist Guide. Oxford University Press, Oxford (2017)

Dayan, P., Huys, Q.J.M.: Serotonin, inhibition, and negative mood. PLoS Comput. Biol. 4(2), e4 (2008). https://doi.org/10.1371/journal.pcbi.0040004

Friston, K., Da Costa, L., Hafner, D., Hesp, C., Parr, T.: Sophisticated Inference (2020). http://arxiv.org/abs/2006.04120

Friston, K.J., Redish, A.D., Gordon, J.A.: Computational nosology and precision psychiatry. Comput. Psychiatr. (Cambridge, Mass.) 1, 2–23 (2017). https://doi.org/10.1162/CPSY_a_00001

Gale, C., Davidson, O.: Generalised anxiety disorder. Br. Med. J. (2007). https://doi.org/10.1136/bmj.39133.559282.BE

Hayes, S.C., Luoma, J.B., Bond, F.W., Masuda, A., Lillis, J.: Acceptance and commitment therapy: model, processes and outcomes. Behav. Res. Ther. 44(1), 1–25 (2006)

Hesp, C., Smith, R., Parr, T., Allen, M., Friston, K., Ramstead, M.: Deeply felt affect: the emergence of valence in deep active inference, 3 December 2019. https://doi.org/10.31234/osf.io/62pfd

Huys, Q.J.M., Eshel, N., O'Nions, E., Sheridan, L., Dayan, P., Roiser, J.P.: Bonsai trees in your head: how the pavlovian system sculpts goal-directed choices by pruning decision trees. PLoS Comput. Biol. 8(3), e1002410 (2012). https://doi.org/10.1371/journal.pcbi.1002410

Huys, Q.J.M., et al.: Interplay of approximate planning strategies. Proc. Nat. Acad. Sci. U.S.A. 112(10), 3098–3103 (2015). https://doi.org/10.1073/pnas.1414219112

Nĭally, N., Huys, Q.J.M., Eshel, N., Faulkner, P., Dayan, P., Roiser, J.P.: The neural basis of aversive pavlovian guidance during planning. J. Neurosci. 37(42), 10215–10229 (2017). https://doi.org/10.1523/JNEUROSCI.0085-17.2017

Causal Blankets: Theory and Algorithmic Framework

Fernando E. Rosas[1,2,3($\boxtimes$)], Pedro A. M. Mediano[4], Martin Biehl[5], Shamil Chandaria[1,6], and Daniel Polani[7]

[1] Centre for Psychedelic Research, Imperial College London, London SW7 2DD, UK
f.rosas@imperial.ac.uk, shamil.chandaria@gmail.com
[2] Data Science Institute, Imperial College London, London SW7 2AZ, UK
[3] Centre for Complexity Science, Imperial College London, London SW7 2AZ, UK
[4] Department of Psychology, University of Cambridge, Cambridge CB2 3EB, UK
pam83@cam.ac.uk
[5] Araya Inc., Tokyo 107-6024, Japan
martin@araya.org
[6] Institute of Philosophy, School of Advanced Study, University of London,
London WC1E 7HU, UK
[7] Department of Computer Science, University of Hertfordshire,
Hatfield AL10 9AB, UK
d.polani@herts.ac.uk

Abstract. We introduce a novel framework to identify perception-action loops (PALOs) directly from data based on the principles of computational mechanics. Our approach is based on the notion of *causal blanket*, which captures sensory and active variables as dynamical sufficient statistics—i.e. as the "differences that make a difference." Furthermore, our theory provides a broadly applicable procedure to construct PALOs that requires neither a steady-state nor Markovian dynamics. Using our theory, we show that every bipartite stochastic process has a causal blanket, but the extent to which this leads to an effective PALO formulation varies depending on the integrated information of the bipartition.

Keywords: Perception-action loops · Computational mechanics · Integrated information · Stochastic processes

1 Introduction

The perception-action loop (PALO) is one of the most important constructs of cognitive science, and plays a fundamental role in many other disciplines including reinforcement learning and computational neuroscience. Despite its importance and pervasiveness, fundamental questions about what kind of systems can

F.R. was supported by the Ad Astra Chandaria foundation. P.M. was funded by the Wellcome Trust (grant no. 210920/Z/18/Z). M.B. was supported by a grant from Templeton World Charity Foundation, Inc. (TWCF). The opinions expressed in this publication are those of the authors and do not necessarily reflect the views of TWCF.

T. Verbelen et al. (Eds.): IWAI 2020, CCIS 1326, pp. 187–198, 2020.
https://doi.org/10.1007/978-3-030-64919-7_19

be properly described by a PALO are still to a large extent unanswered. The aim of this paper is to introduce a framework that allows us to identify PALOs directly from data, which complements existent approaches and serves to deepen our understanding of the essential elements that make a PALO.

1.1 Markov Blankets

One of the most encompassing accounts of PALOs can be found in the Free Energy Principle (FEP) literature, which formalises them via *Markov blankets* (MBs) [14]. An interesting contribution of this literature is to characterise "sensory" (S) and "active" (A) variables as having two defining properties: (i) they mediate the interactions between internal variables of the agent (M) and external variables of its environment (E), and (ii) they impose a specific causal structure on these interactions—e.g. sensory variables may affect internal variables, but are not (directly) affected by them [14].

Formally, MBs were originally introduced by Pearl [21] for Markov and Bayesian networks. Within the FEP literature, MBs are usually employed in multivariate stochastic processes with ergodic Markovian dynamics, with a steady-state distribution p^* that is required to satisfy [20] (Fig. 1a)

$$p^*(e_t, m_t | s_t, a_t) = p^*(e_t | s_t, a_t) p^*(m_t | s_t, a_t) . \tag{1}$$

However, Eq. (1) does not suffice to guarantee a PALO structure, as noted in Ref. [7]. In effect, the MB condition is insufficient to establish requirement (ii): its symmetry with respect to internal and external variables makes it impossible to infer the direction of the loop; additionally, the fact that the condition holds across variables synchronously makes it unsuitable to guarantee a causal relationship [22]. Recent reports [11] acknowledge that this synchronous condition needs to be complemented with additional diachronic restrictions on the system's dynamics, which can be written, for instance, as a set of coupled stochastic differential equations of the form (Fig. 1b)

$$\begin{aligned}
\dot{m}_t &= f_{in}(m_t, a_t, s_t) + \omega_t^{in} , & \dot{a}_t &= f_a(m_t, a_t, s_t) + \omega_t^a , \\
\dot{e}_t &= f_{ex}(e_t, a_t, s_t) + \omega_t^{ex} , & \dot{s}_t &= f_s(e_t, a_t, s_t) + \omega_t^s .
\end{aligned} \tag{2}$$

Above, the functions f_{in}, f_a, f_{ex}, f_s determine the flow, and $\omega_t^{in}, \omega_t^a, \omega_t^{ex}, \omega_t^s$ denote additive Gaussian noise. Interestingly, it has been shown that Eq. (2) implies Eq. (1) under additional assumptions: either block diagonality conditions over the solenoidal flow [11], or strong dissipation [12, Appendix].[1] Hence, PALOs could be interpreted as coupled stochastic dynamical systems of the form in Eq. (2), as long as the flow satisfies any of the two mentioned conditions.

Despite its elegance, this formalisation of PALOs has important limitations. First, this formulation relies strongly on Langevin dynamics, making it difficult to extend it to PALOs appearing in discrete systems. Secondly, this approach

---

[1] However, in the general case neither Eqs. (1) or (2) imply each other [7]—hence they need to be taken as complementary conditions.

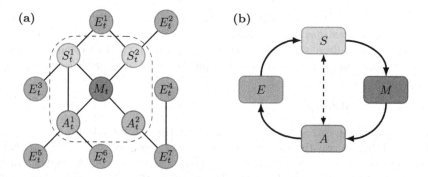

Fig. 1. Two visualisations of PALOs in the FEP literature, either based on **(a)** Markov blankets according to Eq. (1) or **(b)** Langevin dynamics following Eq. (2).

depends on a set of assumptions—for one, the aforementioned conditions over the flow and the restriction to systems in their steady-state—that might be too restrictive for some scenarios of interest. Finally, and perhaps most importantly, Eq. (1) forces all interactions between M_t and E_t to be accountable by (S_t, A_t), which imposes—due to the data processing inequality [9]—an information bottleneck of the form $I(M_t; E_t) \leq I(M_t; A_t, S_t)$. Therefore, the MB formalism forbids interdependencies induced by past events that are kept in memory, but may not directly influence the present state of the blankets.[2] This information kept in memory arguably plays an important role in many PALOs, and includes uncontroversial features of cognition (such as old memories that an agent retains but is neither caused by a sensation nor causing an action at the current moment), yet are forbidden by MBs.

1.2   Computational Mechanics, Causal States, and Epsilon-Machines

Computational mechanics is a method for studying patterns and statistical regularities observed in stochastic processes by uncovering their hidden causal structure [24,25]. A key insight is that an optimal, minimimal representation of a process can be revealed by grouping past trajectories according to their forecasting abilities into so-called *causal states*. More precisely, the causal states of a (possibly non-Markovian) time series $\{Z_t\}_{t \in \mathbb{Z}}$ are the equivalent classes of trajectories $\overleftarrow{z}_t := (\ldots, z_{t-1}, z_t)$ given by the relationship

$$\overleftarrow{z}_t \equiv_\epsilon \overleftarrow{z}_t' \qquad \text{iff} \qquad p(z_{t+1}|\overleftarrow{z}_t) = p(z_{t+1}|\overleftarrow{z}_t') \quad \forall z_{t+1} \, .$$

It can be shown that the causal states are the coarsest coarse-graining of past trajectories $\overleftarrow{x}_t$ that retains full predictive power over future variables [10,13]. Moreover, the corresponding process over causal states always has Markovian dynamics, providing the simplest yet encompassing representation of the system's information dynamics on a latent space—known as the *epsilon-machine*.

---

[2] We thank Nathaniel Virgo for first noting this issue.

Please note that the causal states of a system are guaranteed to provide counterfactual relationships [22] only if the system at hand is fully observed. In the case of partially observed scenarios, causal states ought to be understood in the Granger sense, i.e. as states of maximal non-mediated predictive ability [8].

1.3 Contribution

In this paper we introduce an operationalisation of PALOs based on *causal blankets* (CB), a construction based on a novel definition of dynamical statistical sufficiency. CB capture properties (i) and (ii) in a single mathematical construction by applying informational constructs directly to dynamical conditions. Moreover, CBs can be constructed with great generality for any bipartite system without imposing further conditions, and hence can be applied to non-ergodic, non-Markovian stochastic processes. This generality allows us to explore novel connections between PALOs and integrated information. In the rest of the manuscript, we:

1) Provide a rigorous definition of CBs (Definition 2); and
2) Show every agent-environment partition has a CB, and thus can be described as a PALO (Proposition 1); although
3) Not all systems are equally well described as a PALO, and this can be quantified via information geometry and integrated information (Sect. 3)—providing a principled measure to distinguish preferable candidates for PALO.[3]

2   Causal Blankets as Informational Boundaries

We consider the perspective of a scientist who repeatedly measures a system composed of two interacting parts X_t and Y_t. We assume that, from these observations, a reliable statistical model of the corresponding discrete-time stochastic process can be built—of which all the resulting marginal and conditional distributions are well-defined. Random variables are denoted by capital letters (e.g. X, Y) and their realisations by lower case letters (e.g. x, y); stochastic processes at discrete times (i.e. time series) are represented as bold letters without subscript $\boldsymbol{X} = \{X_t\}_{t \in \mathbb{Z}}$, and $\bar{\boldsymbol{X}}_t := (\dots, X_{t-1}, X_t)$ denotes the infinite past of $\boldsymbol{X}$ until and including t.

Given two random variables X and Y, a statistic $U = f(X)$ is said to be *Bayesian sufficient of X w.r.t. Y* if $X \perp\!\!\!\perp Y \mid U$, which implies that all the common variability between X and Y is accounted for by U [9]. The first step in our construction is to introduce a dynamical version of statistical sufficiency.

---

[3] The proofs of our results can be found in the Appendix.

Definition 1 (D-BaSS). *Given two stochastic processes X, Y, a process U is a dynamical Bayesian sufficient statistic (D-BaSS) of X w.r.t. Y if, for all $t \in \mathbb{Z}$, the following conditions hold:*

i. Precedence: there exists a function $F(\cdot)$ such that $U_t = F(\overset{\leftarrow}{X}_t)$ for all $t \in \mathbb{Z}$.
ii. Sufficiency: $Y_{t+1} \perp\!\!\!\perp \overset{\leftarrow}{X}_t \mid (U_t, \overset{\leftarrow}{Y}_t)$.

Moreover, a stochastic process M is a minimal D-BaSS of X with respect to Y if it is itself a D-BaSS and for any D-BaSS U there exists a function $f(\cdot)$ such that $f(U_t) = M_t, \forall t \in \mathbb{Z}$.

The first condition above states that U is no more than a simpler, coarse-grained representation of X, and the second implies that the influence of $\overset{\leftarrow}{X}_t$ on Y_{t+1} given $\overset{\leftarrow}{Y}_t$ is fully mediated by U_t. This has interesting consequences for transfer entropy, as seen in the next lemma.

Lemma 1. *If U is a D-BaSS of X w.r.t. Y, then*

$$\mathrm{TE}(X \to Y)_t := I(\overset{\leftarrow}{X}_t; Y_{t+1} | \overset{\leftarrow}{Y}_t) = I(U_t; Y_{t+1} | \overset{\leftarrow}{Y}_t) \ . \tag{3}$$

There are many such D-BaSS; e.g. $U_t = \overset{\leftarrow}{X}_t$ would be one valid D-BaSS of X w.r.t. Y. However, Theorem 1 shows that minimal D-BaSS's are unique (up to bijective transformations).

Theorem 1 (Existence and uniqueness of the minimal D-BaSS). *Given stochastic processes X, Y, the minimal D-BaSS of X w.r.t. Y corresponds to the partition of past-trajectories $\overset{\leftarrow}{x}_t$ induced by the following equivalence relationship:*

$$\overset{\leftarrow}{x}_t \equiv_p \overset{\leftarrow}{x}_t' \qquad \textit{iff} \qquad \forall \overset{\leftarrow}{y}_t, y_{t+1} \quad p(y_{t+1} | \overset{\leftarrow}{x}_t, \overset{\leftarrow}{y}_t) = p(y_{t+1} | \overset{\leftarrow}{x}_t', \overset{\leftarrow}{y}_t) \ .$$

Therefore, the minimal D-BaSS is always well-defined, and is unique up to an isomorphism.

This result shows that D-BaSSs can be built irrespective of any other possibly latent influences on X and Y, as it is defined purely on the joint statistics of these two processes. Moreover, Theorem 1 provides a recipe to build a D-BaSS: group together all the past trajectories that lead to the same predictions, which is a key principle of computational mechanics [10,13,24,25]. Therefore, a minimal D-BaSS distinguishes only "differences that make a difference" for the future dynamics, generalising the construction presented in Ref. [6, Definition 1] for Markovian dynamical systems, and being closely related to the notion of sensory equivalence presented in Ref. [3]. With these ideas at hand, we can formulate our definition of causal blanket.

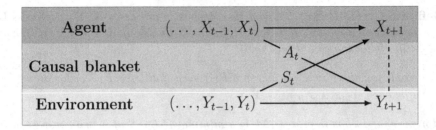

Fig. 2. Causal blanket $\{S, A\}$, which acts as a sufficient statistic mediating the interactions between X and Y.

Definition 2 (Causal blanket). *Given two stochastic processes X, Y, a reciprocal D-BaSS (ReD-BaSS) is a stochastic process R which satisfies:*

i. *Joint precedence: $R_t = F(\breve{X}_t, \breve{Y}_t)$ for some function $F(\cdot)$.*
ii. *Reciprocal sufficiency: R is a D-BaSS of X w.r.t. Y, and also is a D-BaSS of Y w.r.t. X.*

A causal blanket (CB) is a minimal ReD-BaSS: a time series M, itself a ReD-BaSS, such that for all ReD-BaSSs R there exists a function $f(\cdot)$ such that $M_t = f(R_t), \forall t \in \mathbb{Z}$.

This definition satisfies the two key desiderata discussed in Sect. 1.1: (i) a CB mediates the interactions that take place between X and Y, and (ii) it assesses causality by focusing on statistical relationships between past and future. From this perspective, CBs are the "informational layer" that causally decouples the agent's and environment's temporal evolution from each other (see Proposition 2). Additionally, our next result guarantees that CBs always exist, and are unique to each bipartite system.

Proposition 1. *Given X, Y, their CB always exists and is unique (up to an isomorphism). Moreover, their CB is isomorphic to a pair $\{S, A\}$, where A is a minimal D-BaSS of X w.r.t. Y, and S is a minimal D-BaSS of Y w.r.t. X.*

Proposition 1 has two important consequences: it guarantees that CBs *always* exist, and that they naturally resemble a PALO—as visualised in Fig. 2. Please note that this type of PALO formalisation has a rich history, being studied in Refs. [4,5] and variations being considered in Refs. [15,16,26]. In contrast, our framework follows Refs. [3,6] and does not assume active and sensory variables as given, but discovers them directly from the data. As a matter of fact, the "sensory" (S) and "active" (A) variables of CBs correspond (due to Definition 2) to minimal sufficient statistics that mediate the interdependencies between the past and future of X and Y. The construction of CBs imposes no requirements on the system's statistics or its structure—beyond the bipartition, holding also for non-ergodic and also non-stationary systems, and systems with non-Markovian dynamics.

It is also possible to build internal and external states M_t, E_t such that $(M_t, A_t) = X_t$ and $(E_t, S_t) = Y_t$ with great generality. This can be done via an orthogonal completion of the phase space; the details of this procedure will be made explicit in a future publication. In this way, CBs can be thought as suggesting implicit "equations of motion" somehow equivalent to Eq. (2), as shown in Fig. 2. However, it is important to remark that this representation does *not* provide counterfactual guarantees for partially observed systems (see Sect. 1.2).

Example 1. Consider a multivariate stochastic process M, A, E, S whose dynamics follows

$$
\begin{aligned}
M_{t+1} &= f_{\text{in}}(M_t, A_t, S_t) + N_{\text{in}}, & A_{t+1} &= f_{\text{a}}(M_t, A_t, S_t) + N_{\text{a}}, \\
E_{t+1} &= f_{\text{ex}}(E_t, A_t, S_t) + N_{\text{ex}}, & S_{t+1} &= f_{\text{s}}(E_t, A_t, S_t) + N_{\text{s}},
\end{aligned}
\tag{4}
$$

with $N_t^{\text{in}}, N_t^{\text{a}}, N_t^{\text{ex}}, N_t^{\text{s}}$ being independent of M_t, A_t, E_t, S_t (note that Eq. 4 corresponds to a discrete-time version of Eq. (2)). Then, by defining $X_t = (M_t, A_t)$ and $Y_t = (E_t, S_t)$, one can show using Definition 2 that that $\{S, A\}$ is the CB of X, Y—as long as the partial derivatives of $f_{\text{in}}, f_{\text{a}}, f_{\text{ex}}, f_{\text{s}}$ with respect to their corresponding arguments are nonzero.

3   Integrated Information Transcends the Blankets

According to Definition 2, CBs don't depend on the joint distribution $p(x_{t+1}, y_{t+1}|\bar{\boldsymbol{x}}_t, \bar{\boldsymbol{y}}_t)$, but only on the marginals $p(x_{t+1}|\bar{\boldsymbol{x}}_t, \bar{\boldsymbol{y}}_t)$ and $p(y_{t+1}|\bar{\boldsymbol{x}}_t, \bar{\boldsymbol{y}}_t)$. Here we study how meaningful the CB (and the description of the system as a PALO) is when the joint process's dynamics are different from the product of these two marginals.

Let us start by introducing the *synergistic coefficient* $\xi_t \in \mathbb{R}$, which is a random variable given by

$$
\xi_t := \log \frac{p(X_{t+1}, Y_{t+1}|\bar{\boldsymbol{X}}_t, \bar{\boldsymbol{Y}}_t)}{p(X_{t+1}|\bar{\boldsymbol{X}}_t, \bar{\boldsymbol{Y}}_t)\, p(Y_{t+1}|\bar{\boldsymbol{X}}_t, \bar{\boldsymbol{Y}}_t)} .
\tag{5}
$$

A process (X, Y) is said to have *factorisable dynamics* if $\xi_t = 0$ a.s. for all $t \in \mathbb{Z}$.

Proposition 2 (Conditional independence of trajectories). *If R is a ReD-BaSS and the dynamics of X, Y is factorisable, then $X \perp\!\!\!\perp Y \mid R$. Thus, such system is perfectly described as a PALO, and R is a MB (in Pearl's sense).*

A direct consequence of this Proposition is that a ReD-BaSS does not guarantee statistical independence of X, Y at the trajectory level in non-factorisable systems. Therefore, in such systems there are interactions between X and Y that are not mediated by the CB. Please note that this is not a weakness of the CB construction—which is optimal in capturing all the directed influences, as shown in Proposition 1. Instead, this result suggests that non-factorisable systems might not be well-suited to be described as a PALO.

To further understand this, let us explore the integrated information in the system (X, Y) using information geometry [19]. For this, consider the manifolds

$$\mathcal{M}_1 = \left\{ q_t : q(x_{t+1}, y_{t+1} | \bar{x}_t, \bar{y}_t) = q(x_{t+1} | \bar{x}_t, \bar{y}_t) q(y_{t+1} | \bar{x}_t, \bar{y}_t) \right\} ,$$
$$\mathcal{M}_2 = \left\{ q_t : q(x_{t+1}, y_{t+1} | \bar{x}_t, \bar{y}_t) = q(x_{t+1} | \bar{x}_t) q(y_{t+1} | \bar{y}_t) \right\} .$$

Manifold $\mathcal{M}_1$ corresponds to all systems with factorisable dynamics, and $\mathcal{M}_2$ to all systems where the dynamics of agent and environment are fully decoupled. The information-geometric projection of an arbitrary system p_t onto $\mathcal{M}_2$,

$$\tilde{\varphi}_t := \min_{q_t \in \mathcal{M}_2} D(p_t \| q_t) , \tag{6}$$

has been proposed as a measure of integrated information [2,18]. Using the Pythagoras theorem [1] together with the fact that $\mathcal{M}_2 \subset \mathcal{M}_1$, one can decompose $\tilde{\varphi}_t$ as

$$\underbrace{\tilde{\varphi}_t}_{D(p_t \| q_t^{(2)})} = \underbrace{\mathbb{E}\{\xi_t\}}_{D(p_t \| q_t^{(1)})} + \underbrace{\left[\mathrm{TE}(A \to Y)_t + \mathrm{TE}(S \to X)_t \right]}_{D(q_t^{(1)} \| q_t^{(2)})} , \tag{7}$$

where $q_t^{(k)} := \arg\min_{q_t \in \mathcal{M}_k} D(p_t \| q_t)$.[4]

This decomposition confirms previous results that showed that integrated information is a construct that combines low-order transfer and high-order synergies [17]. Thanks to Lemma 1, Eq. (7) states that the transfer component of $\tilde{\varphi}_t$ (i.e. $D(q_t^{(1)} \| q_t^{(2)})$) is what is properly mediated by the CB. In contrast, the part of $\tilde{\varphi}$ related to high-order statistics, i.e. $\mathbb{E}\{\xi_t\} = I(X_{t+1}; Y_{t+1} | \bar{X}_t, \bar{Y}_t)$, is not accounted by the CB. This last term can either refer to spurious synchronous correlations (due e.g. to sub-sampling), or be due to synergistic dynamics that are a signature of emergent phenomena [23].

In summary, our results suggest that the dynamics of a system (X, Y) that is too synergistically integrated are poorly represented as a PALO, even if the CB formally still exists. Additionally, the synergistic component of integrated information can be used as a measure for this mismatch.

4   Conclusion

This manuscript introduced a data-driven method to build PALOs leveraging principles of computational mechanics. Our construction provides an informational interpretation of sensory and actuation variables: sensory (resp. active) variables encode all the changes from "outside" (resp. "inside") that affect the future evolution of the "inside" (resp. "outside"). Our framework is broadly applicable, depending only on the underlying bipartition but not imposing

---

[4] Note that in non-ergodic scenarios the expected values are not calculated over individual trajectories, but over the ensemble statistics that define the probability.

any further conditions on the system's dynamics or distribution. Furthermore, we illustrated how this construction allows one to relate—within a PALO framework—the separation of a system and its environment to the integrated information encompassing the two.

It is to be noted that the CB construction relies on discrete time, which, while being immediately applicable to digitally sampled data, might not be natural in some scenarios. Also, CB theory at this stage does not provide explicit links with probabilistic inference. As shown in Example 1, CBs provide a natural extension of Eq. (2) to the discrete-time case, so one possibility would be to combine them with the MB condition in Eq. (1). The exploration of such "causal Markov blankets" which would satisfy both Eq. (1) and Definition 2 is an interesting avenue for future research.

It is our hope that the CB construction may enrich the toolbox of researchers studying PALOs and help to illuminate further our understanding of the nature of agency.

A   Proofs

Proof (Lemma 1). Let's consider U to be a D-BaSS of X w.r.t. Y. Then, property (ii) of a D-Bass is equivalent to

$$I(\overleftarrow{X}_t; Y_{t+1} \mid U_t, \overleftarrow{Y}_t) = 0 . \tag{8}$$

Using this, one can verify that

$$I(\overleftarrow{X}_t; Y_{t+1}|\overleftarrow{Y}_t) = I(U_t, \overleftarrow{X}_t; Y_{t+1}|\overleftarrow{Y}_t) = I(U_t; Y_{t+1}|\overleftarrow{Y}_t) .$$

Here, the first equality holds because U_t is a deterministic function of $\overleftarrow{X}_t$, and the second equality follows from an application of the chain rule and Eq. (8).

Proof (Theorem 1). Consider the function $F(\cdot)$ that maps each $\overleftarrow{x}_t$ to its corresponding equivalence class $F(\overleftarrow{x}_t)$ established by the equivalence relationship $\equiv_p$, and define $M_t = F(\overleftarrow{X}_t)$. As this construction satisfies the requirement of precedence in Definition 1, let us show the sufficiency of M. By definition of M_t, it is clear that if $m_t = F(\overleftarrow{x}_t)$ then

$$p(y_{t+1}|\overleftarrow{x}_t, \overleftarrow{y}_t) = p(y_{t+1}|m_t, \overleftarrow{y}_t) ,$$

which implies that $H(Y_{t+1}|\overleftarrow{X}_t, \overleftarrow{Y}_t) = H(Y_{t+1}|M_t, \overleftarrow{Y}_t)$. As a consequence,

$$\begin{aligned}
I(\overleftarrow{X}_t; Y_{t+1}|\overleftarrow{Y}_t) &= H(Y_{t+1}|\overleftarrow{Y}_t) - H(Y_{t+1}|\overleftarrow{X}_t, \overleftarrow{Y}_t) \\
&= H(Y_{t+1}|\overleftarrow{Y}_t) - H(Y_{t+1}|M_t, \overleftarrow{Y}_t) \\
&= I(M_t; Y_{t+1}|\overleftarrow{Y}_t) .
\end{aligned} \tag{9}$$

From this, sufficiency follows from noticing that

$$I(\breve{\boldsymbol{X}}_t; Y_{t+1}|M_t, \breve{\boldsymbol{Y}}_t) = I(\breve{\boldsymbol{X}}_t, M_t; Y_{t+1}|\breve{\boldsymbol{Y}}_t) - I(M_t; Y_{t+1}|\breve{\boldsymbol{Y}}_t)$$
$$= I(\breve{\boldsymbol{X}}_t; Y_{t+1}|\breve{\boldsymbol{Y}}_t) - I(M_t; Y_{t+1}|\breve{\boldsymbol{Y}}_t)$$
$$= 0 .$$

Above, the first equality is due to the chain rule, the second follows from the fact that M_t is a function of $\breve{\boldsymbol{X}}_t$, and the third uses Eq. (9).

To finish the proof, let us show that M is minimal. For this, consider another U to be another D-BaSS of X w.r.t. Y. As $U_t = G(\breve{\boldsymbol{X}}_t)$ for some function $G(\cdot)$, U corresponds to another partition of the trajectories $\breve{\boldsymbol{x}}_t$. If there exists no function f such that $f(U_t) = M_t$, that implies that the partition that corresponds to M is not a coarsening of the partition for U, and therefore that there exists $\breve{\boldsymbol{x}}_t$ and $\breve{\boldsymbol{x}}_t'$ such that $G(\breve{\boldsymbol{x}}_t) = G(\breve{\boldsymbol{x}}_t')$ while $p(y_{t+1}|\breve{\boldsymbol{x}}_t, \breve{\boldsymbol{y}}_t) \neq p(y_{t+1}|\breve{\boldsymbol{x}}_t', \breve{\boldsymbol{y}}_t)$. This, in turn, implies that there exists a $\breve{\boldsymbol{x}}_t'$ such that that $p(y_{t+1}|u_t, \breve{\boldsymbol{x}}_t', \breve{\boldsymbol{y}}_t) \neq p(y_{t+1}|u_t, \breve{\boldsymbol{y}}_t) = \sum_{\breve{\boldsymbol{x}}_t} p(y_{t+1}|u_t, \breve{\boldsymbol{x}}_t, \breve{\boldsymbol{y}}_t)p(\breve{\boldsymbol{x}}_t|u_t, \breve{\boldsymbol{y}}_t)$, showing that $\breve{\boldsymbol{X}}_t$ is not conditionally independent of Y_{t+1} given $U_t, \breve{\boldsymbol{Y}}_t$, contradicting the fact that U is a D-BaSS. This contradiction proves that the partition induced by U is a refinement of the partition induced by M, proving the minimality of the latter.

Proof (Proposition 1). Let's denote by A the minimal D-BaSS of X w.r.t. Y, and S the minimal D-BaSS of Y w.r.t. X, which are known to exist and be unique thanks to Theorem 1. Then, by defining $M_t := (S_t, A_t)$, one can directly verify that M is a ReD-BaSS of (X, Y). To prove its minimality, let us consider another ReD-BaSS of (X, Y) denoted by N. As N is a D-BaSS of X w.r.t. Y, the minimality of A guarantees the existance of a mapping $f(\cdot)$ such that $f(N_t) = S_t$. Similarly, thanks to the minimality of S, there is another mapping $g(\cdot)$ such that $g(N_t) = A_t$. Therefore, the function $F(\cdot) = (f, g)$ satisfies $F(N_t) = M_t$, which confirms the minimality of M.

Proof (Proposition 2). The proof is based on the principle that if $p(A, B, C) = f(A, C)g(B, C)$, then $A \perp\!\!\!\perp B|C$. Building on that rationale, a direct calculation shows that

$$p(\boldsymbol{x}, \boldsymbol{y}) = \prod_{\tau=-\infty}^{\infty} p(x_{\tau+1}, y_{\tau+1}|\breve{\boldsymbol{x}}_\tau, \breve{\boldsymbol{y}}_\tau)$$

$$= \prod_{\tau=-\infty}^{\infty} \exp\{\xi_\tau\} \, p(x_{\tau+1}|\breve{\boldsymbol{x}}_\tau, \breve{\boldsymbol{y}}_\tau) \, p(y_{\tau+1}|\breve{\boldsymbol{x}}_\tau, \breve{\boldsymbol{y}}_\tau), \tag{10}$$

where the second equality[5] uses Eq. (5). Additionally, if, as per assumption of the Proposition, R is a ReD-BaSS of (X, Y), then

$$p(x_{\tau+1}|\breve{\boldsymbol{x}}_\tau, \breve{\boldsymbol{y}}_\tau) = p(x_{\tau+1}|\breve{\boldsymbol{x}}_\tau, \breve{\boldsymbol{y}}_\tau, r_\tau) = p(x_{\tau+1}|\breve{\boldsymbol{x}}_\tau, r_\tau),$$

---

[5] Note that the infinite products in this proof are just a formal procedure to acknowledge products that can be taken up to arbitrary times.

where the first equality uses the fact that r_τ (by definition) is a function of $(\bar{\boldsymbol{x}}_\tau, \bar{\boldsymbol{y}}_\tau)$, and the second uses the sufficiency of D-BaSS's. Following an analogous derivation, one can show that $p(y_{\tau+1}|\bar{\boldsymbol{x}}_\tau, \bar{\boldsymbol{y}}_\tau) = p(y_{\tau+1}|r_\tau, \bar{\boldsymbol{y}}_\tau)$. Then, with the assumption that the dynamics of $(\boldsymbol{X}, \boldsymbol{Y})$ is factorisable and hence $\xi_t = 0$, it follows from Eq. (10) that

$$p(\boldsymbol{x}, \boldsymbol{y}) = \prod_{\tau=-\infty}^{\infty} p(x_{\tau+1}|r_\tau, \bar{\boldsymbol{y}}_\tau)\, p(y_{\tau+1}|r_\tau, \bar{\boldsymbol{y}}_\tau)\,.$$

Separating the two product series, this shows that there exist functions $f(\cdot)$ and $g(\cdot)$ such that $p(\boldsymbol{x}, \boldsymbol{y}) = f(\boldsymbol{x}, \boldsymbol{r})g(\boldsymbol{y}, \boldsymbol{r})$, and hence one has $\boldsymbol{X} \perp\!\!\!\perp \boldsymbol{Y}|\boldsymbol{R}$, which completes the proof.

References

1. Amari, S.i., Nagaoka, H.: Methods of Information Geometry, vol. 191. American Mathematical Soc. (2007)
2. Ay, N.: Information geometry on complexity and stochastic interaction. Entropy **17**(4), 2432–2458 (2015)
3. Ay, N., Löhr, W.: The Umwelt of an embodied agent—a measure-theoretic definition. Theory Biosci. **134**(3–4), 105–116 (2015)
4. Bertschinger, N., Olbrich, E., Ay, N., Jost, J.: Information and closure in systems theory. In: Explorations in the Complexity of Possible Life. Proceedings of the 7th German Workshop of Artificial Life, pp. 9–21 (2006)
5. Bertschinger, N., Olbrich, E., Ay, N., Jost, J.: Autonomy: an information theoretic perspective. Biosystems **91**(2), 331–345 (2008)
6. Biehl, M., Polani, D.: Action and perception for spatiotemporal patterns. In: Artificial Life Conference Proceedings, vol. 14, pp. 68–75. MIT Press (2017)
7. Biehl, M., Pollock, F.A., Kanai, R.: A technical critique of the free energy principle as presented in "Life as we know". arXiv:2001.06408 (2020)
8. Bressler, S.L., Seth, A.K.: Wiener-Granger causality: a well established methodology. Neuroimage **58**(2), 323–329 (2011)
9. Cover, T.M., Thomas, J.A.: Elements of Information Theory. Wiley, Hoboken (2012)
10. Crutchfield, J.P., Young, K.: Inferring statistical complexity. Phys. Rev. Lett. **63**(2), 105 (1989)
11. Friston, K., Da Costa, L., Parr, T.: Some interesting observations on the free energy principle. arXiv:2002.04501 (2020)
12. Friston, K.J., et al.: Parcels and particles: Markov blankets in the brain. arXiv:2007.09704 (2020)
13. Grassberger, P.: Toward a quantitative theory of self-generated complexity. Int. J. Theor. Phys. **25**(9), 907–938 (1986)
14. Kirchhoff, M., Parr, T., Palacios, E., Friston, K., Kiverstein, J.: The Markov blankets of life: autonomy, active inference and the free energy principle. J. R. Soc. Interface **15**(138), 20170792 (2018)
15. Klyubin, A.S., Polani, D., Nehaniv, C.L.: Organization of the information flow in the perception-action loop of evolved agents. In: Proceedings. 2004 NASA/DoD Conference on Evolvable Hardware, pp. 177–180. IEEE (2004)

16. Klyubin, A.S., Polani, D., Nehaniv, C.L.: Representations of space and time in the maximization of information flow in the perception-action loop. Neural Comput. **19**(9), 2387–2432 (2007)
17. Mediano, P.A., Rosas, F., Carhart-Harris, R.L., Seth, A.K., Barrett, A.B.: Beyond integrated information: a taxonomy of information dynamics phenomena. arXiv:1909.02297 (2019)
18. Mediano, P.A., Seth, A.K., Barrett, A.B.: Measuring integrated information: comparison of candidate measures in theory and simulation. Entropy **21**(1), 17 (2019)
19. Oizumi, M., Tsuchiya, N., Amari, S.i.: Unified framework for information integration based on information geometry. Proc. Nat. Acad. Sci. **113**(51), 14817–14822 (2016)
20. Parr, T., Da Costa, L., Friston, K.: Markov blankets, information geometry and stochastic thermodynamics. Philos. Trans. Roy. Soc. A Math. Phys. Eng. Sci. **378**(2164), 20190159 (2020)
21. Pearl, J.: Probabilistic Reasoning in Intelligent Systems: Networks of Plausible Inference. Morgan Kaufmann, Burlington (1988)
22. Pearl, J.: Causality. Cambridge University Press, Cambridge (2009)
23. Rosas, F.E., et al.: Reconciling emergences: an information-theoretic approach to identify causal emergence in multivariate data. arXiv:2004.08220 (2020)
24. Shalizi, C.R., Crutchfield, J.P.: Computational mechanics: pattern and prediction, structure and simplicity. J. Stat. Phys. **104**(3–4), 817–879 (2001)
25. Shalizi, C.: Causal architecture. Complexity, and self-organization in time series and cellular automata. Ph.D. thesis, University of Wisconsin-Madison, Madison, WI (2001)
26. Tishby, N., Polani, D.: Information theory of decisions and actions. In: Cutsuridis, V., Hussain, A., Taylor, J. (eds.) Perception-Action Cycle, pp. 601–636. Springer, New York (2011). https://doi.org/10.1007/978-1-4419-1452-1_19

Author Index

Printed in the United States
By Bookmasters